An Audubon Handbook

A Chanticleer Press Edition

McGraw Hill

Eastern Birds

John Farrand, Jr.

McGraw-Hill Book Company
New York   St. Louis   San Francisco   Auckland   Bogotá
London   Mexico   Montreal   New Delhi   Panama   São Paulo
Singapore   Sydney   Toronto

First Printing

ISBN: 0-07-019976-0

Library of Congress Cataloging-in-Publication Data
Farrand, John.
Eastern birds.
(An Audubon handbook)
Includes index.
1. Birds—Atlantic States—Identification.
2. Birds—Middle West—Identification.   I. Title.
II. Series
QL683.A87F37   1988   598.2974   87-3430
ISBN 0-07-019976-0

Prepared and produced by Chanticleer Press, Inc., New York

Color reproductions made in Italy
Printed and bound in Japan
Typeset by Dix Type Inc., Syracuse, New York

Trademark "Audubon Society" used by publisher under license from the National Audubon Society, Inc.

Jacket photographs of American Black Duck by Rod Planck; title photograph of Forster's Terns by Chuck Gordon; photograph of White-tailed Hawk, pages 18–19, by Larry R. Ditto; photograph of American Coot chasing Blue-winged Teal, pages 478–479, by Gary R. Zahn.

Chanticleer Staff
Publisher: Paul Steiner
Editor-in-Chief: Gudrun Buettner
Executive Editor: Susan Costello
Managing Editor: Jane Opper
Senior Editor: Ann H. Whitman
Associate Editor: David Allen
Assistant Editor: Eva Colin
Production Manager: Helga Lose
Production Assistants: Gina Stead-Thomas,
Philip Rappaport
Art Director: Carol Nehring
Art Associate: Ayn Svoboda
Art Assistant: Cheryl Miller
Picture Library: Edward Douglas

Design: Massimo Vignelli

Acknowledgments and Author

Acknowledgments

In the preparation of this field guide, I am greatly indebted to Gudrun Buettner for her early conception of a handbook on how to identify birds, with two coordinated field guides based on the methods explained in the handbook. I much appreciate her encouragement, inspiration, and invaluable suggestions. To Susan Costello and Jane Opper, I am most grateful for their continuing ideas and advice as they closely followed every step in the preparation of these volumes.

Both this guide and its companion guide to western birds have benefited greatly from careful editing by Ann H. Whitman, who also wrote some of the species accounts. I would like to thank David Allen, who enthusiastically shepherded the species accounts and other text pieces through their many stages. I am very grateful to Carol Nehring for her skillful work with design and graphic matters; to Helga Lose for her expertise in seeing these books through the intricate steps of their production; to Edward Douglas for his assistance in gathering the thousands of photographs; and to Karel Birnbaum and Timothy Allan for their help in editorial matters. Finally, I would like to express my deep appreciation to Paul Steiner for his support during the preparation of this book.

For contributing valuable information to the species accounts, I would especially like to thank Henry T. Armistead, Stephen F. Bailey, Louis R. Bevier, Wayne R. Petersen, Elizabeth C. Pierson, Jan E. Pierson, H. Douglas Pratt, Scott B. Terrill, and Richard E. Webster. I am indebted as well to Michael Lee Bierly, Susan Roney Drennan, Kenn Kaufman, and Kenneth C. Parkes for valuable advice during the preparation of this field guide, and to Les Line, Editor-in-Chief of *Audubon* magazine, for his constant encouragement.

Lastly, for their photographs, patience, and consideration, I am very grateful to Thomas W. Martin, Anthony Mercieca, Wayne Lankinen, and the 165 other photographers, whose work so enhances this field guide.

About the Author

John Farrand, Jr., received a Master of Science degree in zoology from Louisiana State University. He has studied birds throughout North America and in the American tropics, Europe, and Africa. Mr. Farrand is a former editor of *American Birds*, the ornithological journal of the National Audubon Society, a past president of the Linnaean Society of New York, an elective member of the American Ornithologists' Union, and a life member of the Wilson Ornithological Society. He is co-author of *The Audubon Society Field Guide to North American Birds (Eastern Region)* and editor of *The Audubon Society Master Guide to Birding*. He has led birding tours in North America as well as in Europe, where he specializes in the natural history of Greece.

Contents

The Audubon Society

The National Audubon Society is among the oldest and largest private conservation organizations in the world. With over 525,000 members and more than 500 local chapters across the country, the Society works in behalf of our natural heritage through environmental education and conservation action. It protects wildlife in more than 70 sanctuaries from coast to coast. It also operates outdoor education centers and ecology workshops, and publishes the prize-winning *Audubon* magazine, *American Birds* magazine, newsletters, films, and other educational materials. For information regarding membership in the Society, write to the National Audubon Society, 950 Third Avenue, New York, New York 10022.

Foreword by Les Line
My Very First Bird

Editor of Audubon *since 1966, Les Line has been the guiding spirit behind this much-acclaimed nature and conservation magazine.*

You could say that I have been a birder since the crib, for the first sound I can remember from childhood is the scolding of the House Wrens that occupied a hollow-log birdhouse hanging from my mother's clothesline, just beyond the bedroom window. Those were the days when wringer-washers were the latest laundry appliance, clothes-driers were as rare as television sets, and the wash by ancient tradition was done on Monday morning and hung in the backyard to be sun-dried like prunes.

Since a pair of wrens returned May after May to reclaim their northern domicile and refill it with a fresh collection of twigs and grass in preparation for raising a new generation, the weekly disturbance from Mom's Monday ritual couldn't have been all that bothersome. But you would never have known it from their reaction. House Wrens are exuberant singers throughout the nesting regime, proclaiming their marital and territorial rights with waterfalls of sweet, liquid notes. But the grating chatter of a House Wren with its dander up— be the reason a feathered, furred, or upright trespasser—has no redeeming musical qualities whatsoever. It is a warning to retreat. Or else!

There are other bird memories from early boyhood. The perfect halves of powder-blue eggshells found discarded a safe distance from the robin's nest in the maple shading our porch swing. The sad cooing of Mourning Doves when I slept late on June mornings after school was out. The Chipping Sparrow's nest discovered in the spiraea hedge. What a wonderful bit of architecture is the home of the friendly "chippy." A cup tightly woven from grass and rootlets and lined with animal hair.

And the Baltimore Orioles, ah the orioles! They were my favorite then, and remain so today. An American elm towered over our yard. And on May afternoons, while my schoolmates were whacking baseballs in the empty lot behind the Methodist church, I would lie on my back and watch the orioles' courtship and nest-building, a chore the male oriole leaves to his mate. It took her a week to build a pouch perhaps eight inches deep and three inches wide, hanging upside down from the twigs supporting the nest and intricately stitching together hair, strips of bark, plant fibers, bits of string and yarn, and adding soft down from dandelions and willows.

There was a long period when birds played no role in my life. For rekindling that childhood spark, I thank friends who were older and wiser than I. For otherwise I would have missed the return, last Sunday, of the orioles to the great sugar maple shading another front lawn far from my first Michigan home. The final strains of a Mozart wind serenade came over the stereo as the male oriole added his flute song to the voices of oboe, clarinet, bassoon, and horn, and I rushed outside with joy in my heart to greet him.

Such are the greatest rewards of bird watching. Not simply "collecting" new species hither and yon like stamps but getting to know everyday birds like good friends. And so I welcome John Farrand's newly conceived handbook to eastern birds, and in particular its companion volume, *How to Identify Birds*, which make entry into this favorite pastime of so many Americans so much easier.

A New Approach to Birding

Many of us have discovered that it is not only fun to observe birds in the wild, but that it is also wonderfully satisfying to identify them. It takes time, however, to get to know the birds of North America and to identify them with the skill of an expert.

The intent of this all-photographic field guide is both to simplify and speed up the identification process. The 1,354 color photographs, 179 black-and-white drawings, 100,000 words of text, and 70 flight comparisons illustrate and describe over 460 species of birds regularly found east of the Rockies. (A second guide, *An Audubon Handbook: Western Birds*, covers species found west of the Rockies.) The wealth of information presented here, featuring the most useful field marks, is specially suited to the identification of birds. In addition, this guide is coordinated with a companion volume entitled *An Audubon Handbook: How to Identify Birds*, which explains my easy-to-learn method for quick and accurate identification of birds. Whether you are a beginner or an experienced birder and already own a field guide or two, this guide will introduce you to a new way of looking at birds.

Geographical Scope

This guide covers birds regularly found east of the Rocky Mountains, that is, the states between the Atlantic Coast and the western Great Plains, as well as provinces along the southern edge of Canada. It includes birds of the lower Rıo Grande Valley in Texas and the birds of the hilly country in the central portion of that state, but it does not cover the birds of western Texas, the Oklahoma Panhandle, extreme western Kansas, Nebraska, the Dakotas, or Saskatchewan. For these and other western species, you should consult the companion field guide, *An Audubon Handbook: Western Birds*. The arctic northern parts of Quebec, Ontario, and Manitoba are beyond the geographical range of these guides. You should also bear in mind that depending on weather conditions and other natural variables, a species usually found outside our range may occur in the East. However, this guide does not cover these very rare migrants.

Arrangement

To help you refer to a species quickly, the picture-and-text accounts are organized according to similarities such as habitat, lookalikes, and related species. Because I have not adhered to a strictly scientific arrangement, no knowledge of taxonomy is required to find a species.

Bird Groups

Preceding the picture-and-text accounts is the section Bird Groups. Here I describe 62 groups of birds. Each group includes members that are very similar to one another and whose resemblance can be seen clearly by a beginner. If you know the field marks that members of a group share, you can speed up identification, because you will narrow down the number of possible species. However, since the species accounts are arranged by similarities, you do not need to learn these groups to use this guide. But if you do know the name of a particular group, you will find it listed in the table of contents and can turn directly to that section.

*Ne'er look for the birds of this year
in the nests of the last.*

Miguel de Cervantes, 1615
Don Quixote

Picture-and-Text Accounts
Concise text accounts are conveniently located on the same page as
the photographs. Each species is illustrated with one or more
pictures, often showing different plumages, postures, or sexes. The
number next to each photograph corresponds to the number in
parentheses in the text, explaining the view shown. In many
instances, you'll find a close-up of a distinguishing field mark, such as
the head or bill. For birds commonly identified in flight, a photograph
or drawing of the bird in flight supplements the other identification
photographs.

Habitat and Size
For easy reference, the most basic field marks—habitat and size—
appear prominently at the top of the page. The habitat statement tells
the most likely places to find the species. To help you gauge the size of
a bird, every species is assigned to one of seven comparative size
categories: Very Small, Sparrow-sized, Robin-sized, Pigeon-sized,
Crow-sized, Goose-sized, and Very Large. A size "yardstick" appears
above the photographs, so you can compare different species at a
glance. This yardstick is divided into seven sections, and the shaded
area indicates the size category. For example, if three sections are
shaded, this represents the third size, Robin-sized.

Field Marks
The text beneath the photographs provides essential details about the
species in three sections: Field Marks, Similar Species, and Range.
The Field Marks section begins with the average body length of the
adult, followed by information on behavior, shape and posture,
breeding, and any specialized or local habitat not listed in the habitat
statement above. The physical description emphasizes color and
pattern field marks for males, females, and immatures, as well as
special features of geographical races or forms. Distinctive calls and
songs are described or a simple phonetic transcription is given for
all species where this is helpful in identification.

Similar Species
This section tells you how to distinguish lookalikes among the birds
that have full picture-and-text accounts in the book, and also provides
descriptions and scientific names of other, less common species you
may encounter.

Range
Here is the normal geographical range of the species. For migratory
birds, the breeding range is given first, followed by the winter range.
Birds that do not migrate are referred to as "resident." Remember
that some species may wander beyond their normal range.

How to Identify Birds Handbook
While this guide can stand on its own, it gains another dimension
when used in conjunction with its companion volume, *An Audubon
Handbook: How to Identify Birds*, which describes and illustrates the
system I use to identify birds. In that book, I analyze the important
features of birds, and present them in six categories of field marks—

A New Approach to Birding
Parts of a Bird

habitat, size, behavior, shape and posture, color and pattern, and voice. Each of these categories is fully explained with hundreds of examples of eastern and western species. A special feature of the handbook is a series of field mark charts keyed to this field guide. Thus, if you were to become familiar with the approach I have described in *How to Identify Birds*, you would benefit further from this field guide, because the two books are coordinated.

Parts of a Bird
In using this field guide and the companion handbook, you need not know a great deal of technical information to identify birds. However, you will find it helpful to learn a few basic terms for the parts of a bird. These are labeled on the accompanying diagram. All of these parts are field marks for various species, and in combination, they form distinctive patterns that will enable you to identify species or groups. For convenience, these parts can be divided into three general categories: the head, the underparts, and the upperparts, including the wings and tail.

Head

The color and pattern on the head of a bird are often giveaway field marks for identification. Several groups of ducks are easy to recognize by their head patterns, and many groups of songbirds have distinctive markings on the head that immediately allow you to single out a species. Among the warblers, for example, nearly every species has a unique head pattern, made up of different combinations of eyebrows, throat patches, mustaches, and masks. Similarly, sparrows can often be recognized by the color of the eyebrows, crown, or ear patch.

Underparts

Consisting of the breast, flanks, belly, and undertail coverts, the underparts provide many field marks. Thrushes are easily recognized by the spotting on the breast, and among the sparrows, some species, such as the Song and Savannah sparrows, have streaked breasts, while others, including the Swamp and Grasshopper sparrows, have plain breasts.

Upperparts

Field marks involving the upperparts are especially useful because they are what you will see if a bird is flying away from you; a contrastingly colored rump, wingbars, and white outer tail feathers are diagnostic features for many species. Among the vireos, for instance, some species have wingbars and others do not; wingbars are useful also in sorting out the smaller flycatchers.

Other Terms

Four terms that are not labeled on the diagram because they only apply to a limited number of species or bird groups may also be helpful in reading some of the text descriptions: cere, axillars, speculum, and tertials. The cere is the fleshy area around the nostrils, seen in hawks, eagles, falcons, pigeons, doves, and the Budgerigar. Axillars are the feathers under the wing where it meets the body;

The parts of a bird labeled on this drawing are important as field marks, not only singly, but in combination as patterns that are helpful in identifying species.

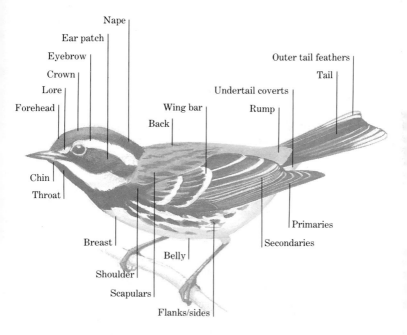

A New Approach to Birding
Feeding, Flocking, and Nesting

these are only important field marks when contrastingly colored in some species. The speculum and tertials refer to the upperparts. Many waterfowl, especially dabbling ducks, have a white or metallic blue, violet, or green area on the secondaries, called the speculum. Tertials are the innermost two or three secondaries; in some species these differ in color or shape from the rest of the secondaries and are therefore useful field marks.

Birds in Nature
Knowing the parts of a bird gives you a vocabulary with which to describe them, but the most important and enjoyable reason for learning what to look for is that this knowledge enables you to observe birds in their natural habitats. It is fascinating to note how birds interact with the plants and animals that make up their environment. Since birds are colorful creatures that rely on sight and sound just as we do, they are probably the most conspicuous part of our wildlife. Although birds are prey for some other animals, including other birds, it is a bird's own search for food that influences its choice of habitat, its seasonal movements, the timing of its nesting, and even its distinctive shape.

Feeding Habits
Just as the food of birds is extremely varied, so are their methods for obtaining it. Fish are captured by many water birds throughout the year. Cormorants and mergansers dive from the surface; gannets, terns, and kingfishers plunge into the water from the air; and herons and egrets stalk the shallows catching fish with their spear-shaped bills. Small mammals and birds are taken by raptors and owls, which seize them with their talons, and by shrikes, which capture their prey with their stout, hooked bills.
Insects form the diet of many species, a majority of which have slender bills. Most insect-eaters breed in spring and early summer, when insects are numerous, and migrate south in winter when it is too cold for their prey to be abroad.
Many seed-eating birds—their hallmark a conical bill—can spend the winter in the North, where seeds remain available throughout the year. A few seed-eating birds, such as the American Goldfinch, nest in midsummer or even later, when the year's crop of seeds is at its peak. Berries and other fruits are important to waxwings, orioles, and certain other birds. Among the fruit-eaters, Cedar Waxwings often breed later in the season, when large numbers of berries are available to feed their young.

Flocking Patterns
Insect-eaters—among them warblers, vireos, and thrushes—are most often seen alone because they must carefully search out prey. In contrast, migrating water birds, commuting between good wetland feeding areas, usually travel in flocks. Similarly, seed-eating and fruit-eating birds usually travel in groups to take full advantage of their unevenly distributed food source, which may be abundant in one place but absent entirely in another.

*Did you ever chance to hear the
midnight flight of birds passing
through the air and darkness
overhead, in countless armies,
changing their early or late
summer habitat?*

Walt Whitman, 1863
Specimen Days

Nesting Requirements

Many birds have very exacting habitat requirements. Not only do
they need abundant food and feeding places suited to their foraging
methods, but also safe nesting sites. Where nesting sites are scarce
but food is abundant or located far from the nesting place, fish-eating
seabirds like terns and auks, long-legged waders like herons, egrets,
and ibises, and aerial insect-hunters like swallows and swifts often
nest in colonies. These birds do not maintain strongly defended
territories. In contrast, raptors, owls, and insect-eaters, which have
difficulty obtaining prey but which can find nesting sites easily, must
stake out a piece of land that is large enough to supply them with food
for themselves and for a brood of hungry nestlings.

The Great Variety of Birds

While birds everywhere have distinct habitat preferences, in the East
the great variety of species you can expect to find is perhaps due to
migration and the changing seasons. A summer woodland is filled with
song, and is inhabited by dozens of species of birds—warblers, vireos,
sparrows, grosbeaks, flycatchers, orioles, and thrushes—while in
winter these same woods are largely silent, populated by a smaller
number of species. Winter bird life includes finches, chickadees,
titmice, nuthatches, woodpeckers, and creepers, whose food supply is
still available. What has happened is an almost complete changeover
of species. The summer birds have departed for warmer places in the
tropics, and have been replaced by another set of species from farther
north—birds that can find food in the more rigorous conditions of
winter. The same replacement can be seen in every habitat—one
group of species in the summer, and a largely different group in the
winter. Between these two seasons are the periods of spring and fall
migration, when dozens of other species that neither breed nor winter
in your area are traveling through on their way between nesting
grounds and warmer winter quarters.

Being aware of this annual cycle can help you to plan your field trips
and to see as large a variety of birds as possible each year. While it is
always a good idea to visit many habitats, you will run up your longest
lists during migration, when many warblers and other songbirds are
passing through woodlands and open-country habitats, and when
great flocks of waterfowl and shorebirds can be found in wetland
habitats. Even in city parks, you'll see dozens of migratory species.
Some of these birds pass through quickly, and you may see them only
once a year if you are lucky. Summer and winter are seasons of
stability, when you will see the same species more frequently and can
get to know them more intimately.

Suburban and Urban Birdlife

We often think of birds as part of the wilderness, or as creatures that
demand unspoiled natural areas. While it is true that most species still
maintain the requirements they had before the arrival of Europeans in
North America, some birds have adapted to a landscape drastically
altered by people. Many common eastern species were originally
open-country birds, absent from much of eastern North America
before the forest was cleared by settlers. Indeed, the original

A New Approach to Birding
Birds in Flight

distribution of many of these species is not clearly known.

Even in the West, which was never subjected to the wholesale destruction of forests that took place in the East, many species can meet all their needs in suburban areas and parks. Here a few acres of shrubs or ornamental trees may harbor jays, warblers, vireos, titmice, woodpeckers, sparrows, and towhees. Food is less varied in cities, but even there, a few species, including not only introduced birds like the Rock Dove, European Starling, and House Sparrow, but also native birds like jays, crows, and the American Robin, find what they need in the way of food and nesting sites, and often have an advantage not enjoyed by birds in wilder places—an absence of predators. It is probably a lack of predators that explains why urban and residential areas, while having fewer species than undisturbed habitats, support more individual birds.

Birds in Flight

Regardless of where you are, you will see birds on the wing. Because of their large size, swans, geese, ducks, raptors, and gulls are among the easiest groups to identify in flight. I have therefore included a special flight section for the most common members of these groups. Flight photographs of 65 species are combined on 12 pages for easy identification and comparison.

A flying bird often appears very different from the way it looks when perching or standing. In flight, a bird may be in view only briefly, giving you just a few seconds to see its field marks. Identifying birds in flight takes sharp eyes and an ability to notice details quickly.

Habitat

It is important, first of all, to note the habitat. Ducks, geese, swans, and gulls are often seen in flight near or over water, although they may also migrate over land. Most raptors, such as eagles, hawks, and falcons, are more likely to be seen in flight over land than over water.

Size

Because it is often impossible to tell how high a bird is flying, judging its size can be difficult. A very useful clue is the speed of its wingbeats. Larger birds tend to have slower or more ponderous wingbeats than smaller ones, and this is often all you need to place a bird in its proper size category. Of course if you already know what group a bird belongs to, you will probably know its size as well.

Behavior

Once you have a clear idea of the size of the bird, you should note its behavior in flight. Is its flight graceful and buoyant, or heavier and more ponderous? Is it soaring—wheeling around as it rides an updraft of warm air? If it is soaring, are its wings flat, or are they held in a shallow V? Is its flight swift and straight, erratic, or bounding? Many birds, especially when migrating or traveling to a roost, fly in bunches or formations. You should note the shape of the formation. All these details are important clues that can help you pin down your identification later.

Shape and Posture
The next set of features to note involves shape. Does the bird have a
long neck or long legs? Are its wings broad and rounded? Or long and
pointed? Is there a distinctive "crook" or "kink" at the bend of the
wing? Is the tail fan-shaped? Or forked? Or pointed? Can you see the
shape of its bill? Noticing these details will help you decide what
broad category—what group—the bird might belong to.

Color and Pattern
In identifying birds in flight, color and pattern are often quite helpful.
The pattern of the wings, wing patches, the color of the underparts,
and the color and pattern of the tail are all features that are usually
easy to see in a flying bird, and these will be useful when you are
deciding what species you have seen.

Voice
Migrating birds often call continuously, as anyone who has seen a V of
Canada Geese knows, and these calls are helpful. So is silence; some
birds, like flocks of gulls, travel past without calling, and this absence
of calls can be as useful as the honking of geese.

Flight Comparison Photographs
On the following pages of flight photographs, the birds are grouped by
their similar appearance—this applies to swans and geese, dabbling
ducks, raptors, and gulls—or by similar habitats—bay ducks or
sea ducks. Important distinguishing features are given in the
accompanying text. Use this section not only to learn the distinctions
among different species, but also to practice seeing the essential field
marks quickly, so that you can apply these techniques to any bird you
may see in flight.

Getting Started
Wherever you are—in the field, your backyard, or even a dense city
or a heavily manicured suburban area—there are birds to be found
and identified. After you have identified a species, the fun is not over,
for then you can begin to watch the bird and see for yourself how it
fits into its environment. You will learn why it chooses the habitats in
which you find it, why it is around only during the summer or winter,
and why you see it nesting or traveling alone. Unraveling these and
other mysteries, both endless and fascinating, are part of what makes
birding so rewarding.

Swans and Geese

You can easily identify swans in flight by their all-white plumage and very long necks. Geese are more varied, but usually show a diagnostic pattern of white and dark. The Snow Goose and Ross' Goose are mainly white, with black wing tips; other species are dark, with white on the face, neck, or tail area.

Tundra Swan

Mute Swan

Ross' Goose

Snow Goose

Brant

Canada Goose

Greater White-fronted Goose

Snow ("Blue") Goose

Male dabbling ducks show a wide range of colors, and are most easily recognized by their shape—slender in the Northern Pintail, stockier in the Mallard, long-tailed in the Wood Duck—and by the white patches on the face, neck, or wings. Females tend to be brown; size, shape, wing pattern, and the male with them, are all helpful clues.

Wood Duck ♂

Mallard ♂

American Wigeon ♂

Northern Pintail ♂

Blue-winged Teal ♂

Cinnamon Teal ♂

Gadwall ♂

American Black Duck

Bay Ducks

The diving ducks that visit bays are stockier than dabbling ducks, and males are patterned in black and white, with few of the bright colors of male dabbling ducks. Note wing patches and stripes, as well as the color and shape of the head. Females are usually brown, with the same wing patterns as males.

Red-breasted Merganser ♂

Ring-necked Duck ♂

Greater Scaup ♂

Lesser Scaup ♂

Canvasback ♂

Common Goldeneye ♂

Bufflehead ♂

Hooded Merganser ♂

Sea Ducks

The males of sea ducks are either all black, black with white patches on the wings or head, or even more boldly patterned in black and white than male bay ducks. Females are duller and usually differently patterned; check shape, especially of the head, and look for light patches on the head or wings.

Harlequin Duck ♂

Oldsquaw winter ♂

King Eider ♂

Common Eider ♂

Surf Scoter ♂

White-winged Scoter ♂

Harlequin Duck ♀

Common Eider ♀

Large Raptors

Shape is the best clue in identifying the largest raptors. Note how rounded or long the wings are, whether the tail is fan-shaped or longer and narrower. The Osprey is easily recognized by the "crook" in its long, rather narrow wings.

Black Vulture

Turkey Vulture

Bald Eagle imm.

Golden Eagle imm.

Flight behavior is also helpful. Most of these birds soar with their wings held flat, but the Turkey Vulture holds its wings in a shallow "V"—or dihedral—and tilts unsteadily from side to side. The Black Vulture flaps its wings much more often than the Turkey Vulture.

Color and pattern are useful, too. The white head and tail of an adult Bald Eagle are all you need to name this bird, and the silvery flash in the wing tips of a Black Vulture is as useful a clue as its rounded wings and frequent flapping. The bold white patch in the wing of a Crested Caracara sets this bird apart from all other raptors.

Golden Eagle

Bald Eagle

Crested Caracara

Osprey

All these species—the Red-tailed Hawk and its relatives—have wide wings and spend much time soaring. There are subtle differences in shape that help in identification, especially since many species, particularly on the Great Plains and in the West, have dark forms in which color clues are not visible.

White-tailed Hawk

Swainson's Hawk

Red-shouldered Hawk

Red-tailed Hawk

All have wide wings, but some have wider and rounder wings than others. The tail may be very broad and fan-shaped or narrower and longer. Some species have relatively large heads. Concentrate on these small differences in shape first. Next, check the color of the tail, the number and width of tail bands, the pattern of the underside of the wing, and the color and pattern of the breast and belly.

Pale areas show in the flight feathers of some species, and a few species have rufous feathers on the legs.

Rough-legged Hawk

Broad-winged Hawk

Snail Kite ♂

Harris' Hawk

Narrow-winged and Short-winged Raptors

In this varied group, the first clue is the shape of the wings—either narrow or short. The falcons, most kites, and the Northern Harrier have narrow wings—pointed in falcons and kites, and blunt-tipped in the Northern Harrier. In contrast, Sharp-shinned Hawks and the larger Cooper's Hawk and Northern Goshawk have short, rounded wings.

Black-shouldered Kite

American Swallow-tailed Kite

Northern Harrier imm.

Peregrine Falcon

*The tails of all these birds are long
and rather narrow, but some are
rounded, others square-tipped.
Size is useful, but it can be hard to
judge in a distant, flying bird, and
you will find shape more helpful.
Details of color can often clinch an
identification.*

Mississippi Kite

Sharp-shinned Hawk

American Kestrel ♀

Merlin ♀

Gulls

Gulls are distinguished from terns by their fan-shaped rather than forked tails, their sturdier, less pointed bills, and their broader wings. Gulls also tend to be larger than terns.

Iceland Gull

Glaucous Gull

Ring-billed Gull winter

Black-legged Kittiwake

Great Black-backed Gull

Franklin's Gull

Bonaparte's Gull

Sabine's Gull

To distinguish adults of the different gull species in flight, you should note the color and pattern of the mantle (the back and upper surface of the wings), the pattern on the upper surface of the wing tips, the amount of blackish coloring under the wings, whether the head is black, and the shape of the wings (narrower in some species than in others). Other helpful clues, harder to see in a flying bird, include the color of the bill and feet, as well as the shape of the bill—stout or slender. Immature gulls may take three to four years to acquire adult plumage, so they are more difficult to identify. Watch for signs of incoming adult plumage, and check the species of nearby adults.

Thayer's Gull winter

Herring Gull

Lesser Black-backed Gull

Laughing Gull br.

Little Gull

Common Black-headed Gull

Bird Groups
Albatrosses to Nightjars

The 62 bird groups described here are informal groups of North American birds whose members clearly resemble one another in behavior, shape and posture, or color and pattern. These groups do not always correspond to families and subfamilies. Not all of these groups are represented in the East.

Albatrosses   Very large seabirds with long, narrow, pointed wings. Often soar in stiff-winged flight, with slow wingbeats; swim and feed at surface.

Anis   Black relatives of cuckoos with flattened, parrotlike bills and long, rounded tails. Flap and coast in loose-winged flight.

Auks   Stocky seabirds with large heads and short wings; some have facial plumes; bill variable, and parrotlike in puffins. Posture often vertical. Swim and dive from surface; fly with rapid wingbeats.

• **Blackbirds**   Mainly black, sparrow- to pigeon-sized birds that walk on ground, fly in bunches, and often mob predators.

• **Bluebirds**   Sparrow-sized thrushes with blue plumage and slender bills. Often perch with tail pointed downward.

Boobies and Gannets   Goose-sized to very large seabirds with stout, spear-shaped bills and long, narrow, pointed wings. Fly in beeline with slow wingbeats; dive for fish from air.

• **Buntings**   Brightly or boldly colored birds with short, conical bills; females greenish or brown.

• **Chickadees**   Sparrow-sized birds with dark crowns, black throats, and white cheek patches. Forage rapidly in foliage and often hang upside down from twigs.

Chickenlike Birds   Stocky, ground-dwelling birds with small heads; often with crest, head or neck plumes, or combs; bill chickenlike; wings broad and rounded. Walk or run; flush with rapid wingbeats. The very large Wild Turkey belongs to this group.

Cormorants   Dark diving birds with slender, hooked bills. Swim with body low. Stand upright; hold wings open to dry after diving. Fly in a line or a V, with long neck extended, flapping and coasting.

• **Crows and Ravens**   Large black birds with stout bills; tail wedge-shaped in ravens. Walk on ground; often travel in bunches and mob predators; ravens soar.

Cuckoos   Slender, robin-sized birds with slightly downcurved bills and long tails. Skulk in vegetation; fly very fast in beeline.

Ducks   Stocky or slender waterfowl, smaller than geese or swans, with flattened bills. Dabbling ducks feed at surface, often tipping up; bay and sea ducks dive from surface. Most fly in formations, on rapid wingbeats; some fly with neck extended or have erratic flight.

• **Finches**   Seed-eating birds; usually with red, pink, or yellow; most have short, conical bills, but crossbills have crossed mandibles for extracting seeds from cones. Fly in bunches, with bounding flight.

• **Flycatchers**   Mostly dully colored, very small to robin-sized birds. Most perch upright and dart out after passing insects; many flick tail; hold tail pointed downward. (Strictly speaking not songbirds, but so closely related in behavior, shape and posture, and color and pattern that the group is included here as songbirds.)

Geese   Long-necked waterfowl with stout bills. Swim and feed at water surface, often tipping up; usually walk on land. Fly in formation with slow wingbeats and necks extended.

Gnatcatchers Very small, slender gray birds with long tails and short, slender bills. Forage rapidly, flicking tail, and cocking tail over back.

Grebes Robin- to crow-sized diving birds that somewhat resemble ducks; some have face or head plumes; most have slender, spear-shaped bill. Seldom fly; often swim with body low in water and dive from surface.

Grosbeaks Sparrow- to robin-sized birds with stout, conical bills. Males brightly colored, females brown or greenish.

Gulls Mainly white water birds with stout, very slightly hooked bills, long, narrow, pointed wings, and usually short, fan-shaped tails. Swim and feed at water surface; mob predators; rob food from other birds; larger species soar.

Hummingbirds Very small to sparrow-sized nectar-feeders with very long and needlelike bills. Usually hover at flowers with very rapid wingbeats and fly in a beeline.

Jaegers and Skuas Dark seabirds with bill gull-like but more hooked; wings long, narrow, pointed. Jaegers have long central tail feathers; skuas have short, fan-shaped tails. Flight erratic, but graceful and swallowlike in smaller species; all rob food from other birds.

Jays Mainly blue birds with stout bills and rounded tails; some have pointed crests. Flight bounding; mob predators.

Juncos Small birds with short, conical bills and white outer tail feathers. Flush when disturbed. Dark-eyed Junco hops; Yellow-eyed Junco walks.

Kingfishers Large-headed birds with bushy crests and stout, spear-shaped bills. Perch or hover over water, then dive to capture fish.

Kinglets Very small, slender-billed birds; olive-green with a bright crown patch. Often flick tails; forage rapidly in foliage, hover, and hang upside down.

Larks Streaked birds with slender bills and white outer tail feathers. Walk or run on ground, flush when disturbed; have bounding flight, and usually fly in bunches.

Long-legged Waders Pigeon-sized to very large wading birds with long necks and long legs. Herons, egrets, and bitterns fly with neck folded over back; others fly with neck extended.

Longspurs Sparrow-sized birds with short, conical bills. Walk on ground; flush when disturbed. Fly in bunches, with bounding flight and simultaneous banking.

Loons Goose-sized or very large water birds with stout, spear-shaped bills and very short tails. Dive from water surface; often swim with body low. Fly in a beeline, with long neck extended.

Magpies Black-and-white birds with stout bills and long, pointed tails. Walk on ground.

Meadowlarks Brown, open-country birds with long, pointed bills. Walk on ground and flush when disturbed.

Nightjars Slender, aerial birds, with large heads, that perch in horizontal posture and catch insects at night or dusk. Flight mothlike (true nightjars) or loose-winged and erratic (nighthawks). Clad in "dead-leaf" patterns; flush from ground.

Bird Groups
Nuthatches to Wrens

Thirteen species found in the East are so distinctive in behavior or shape and posture that they form a "group" of one. These are not included here, because you will find them easily by name in the index.

- **Nuthatches** Stocky, short-necked birds with long, pointed bills and short tails. Scramble up or down trees in search of insects.

- **Orioles** Black-and-orange or black-and-yellow birds that forage slowly in trees, sometimes feed at flowers.

Owls Stocky, mainly nocturnal birds of prey with vertical posture; large, rounded heads, often with ear tufts. Most have camouflage colors. A few hunt during the day.

Pelicans Very large water birds with huge bills. Fly in formations with neck folded; soar, flap, and coast, with slow wingbeats.

Pigeons and Doves Small-headed birds with fan-shaped or pointed tails. Walk on ground; fly very fast in beeline, with rapid wingbeats; often fly in bunches, sometimes with simultaneous banking.

- **Pipits** Sparrow-sized, streaked birds with short, slender bills and white outer tail feathers, usually seen in open country or alpine tundra. Walk or run on ground, flush, and often fly in bunches.

Rails Brown-streaked marsh birds with small head; bill long and slim or chickenlike; wings short, broad, and rounded; tail very short. Walk, run, or wade; probe in mud; flush from grass and fly with rapid wingbeats.

Raptors Hook-billed, predatory birds, often with vertical posture. Head naked in vultures. Most raptors have long, broad, rounded wings, but wings narrow and pointed in falcons and some kites; tail usually fan-shaped or long and rounded. Often soar or flap and coast, some with wings in dihedral.

Shearwaters Seabirds with long, narrow, pointed wings and short, fan-shaped tails. Flap and coast in stiff-winged flight, often in bunches; swim and feed at surface.

Shorebirds A varied group of birds usually seen on or near water; they have long, narrow, pointed wings; larger species have long necks; bill usually short and slender, but in some downcurved, upcurved, or long and straight. Walk or run, and often wade; a few bob head or tail; fly with rapid wingbeats, often in bunches or formations, with simultaneous banking. A few species live chiefly on land.

- **Shrikes** Large-headed, predatory birds with horizontal posture and stout, hooked bill; gray with black mask. Swoop up to perch when landing.

- **Sparrows** Brown-streaked birds with short, conical bills. Skulk in vegetation; forage on ground and flush when disturbed; some scratch noisily in leaves.

Storm-petrels Blackish or gray sparrow- to robin-sized seabirds; tail squared, shallowly forked, or wedge-shaped; some have white rump. Flight swallowlike; hover at water surface to feed.

- **Swallows** Highly aerial sparrow- to robin-sized birds with long, narrow, pointed wings and very small bill. Flight graceful, erratic, and buoyant.

Swans Very large, long-necked white waterfowl with long, gooselike bills. Swim and feed at surface, often tipping up. Fly in formation with slow wingbeats, neck extended.

Anhinga
Budgerigar
• Red-whiskered Bulbul
• Gray Catbird
• Brown Creeper
• Dickcissel
Magnificent Frigatebird
Northern Jacana

• Northern Mockingbird
Greater Roadrunner
Black Skimmer
• European Starling
• Verdin

Swifts   Highly aerial sparrow-sized birds with long, narrow, pointed wings and very short bills. Flight erratic; wingbeats rapid. Wings held more stiffly and beat more rapidly than those of the swallows.

• **Tanagers**   Sparrow- or robin-sized birds with stout but not conical bills. Usually forage in trees. Males brightly colored, females olive or yellowish.

Terns   Slender water birds with horizontal posture; bill slim and pointed or stout and spear-shaped; some have bushy crest; wings long, narrow, pointed; tail usually forked. Flight swallowlike; hover over water and dive from air; mob predators.

• **Thrashers**   Slender birds, mostly with downcurved bills and long tails. Walk or run on ground, scratch noisily in leaves, and skulk in thickets.

• **Thrushes**   Slender-billed, brown birds that often forage on ground, flipping over leaves.

• **Titmice**   Sparrow-sized gray birds with pointed crest, large head, very small bills. Forage rapidly in foliage, and hang upside down from twigs.

• **Towhees**   Ground-dwelling birds with short, conical bills and long, usually rounded tails. Skulk and, when disturbed, flush; scratch noisily in leaves.

Tropicbirds   Mainly white seabirds with stout, spear-shaped bills, long, narrow pointed wings, and very long central tail feathers. Dive for fish from air.

• **Vireos**   Mainly olive, very small to sparrow-sized birds with short, slender bills. Forage slowly for insects in foliage. Several have white wing bars, eye-rings, or spectacles.

• **Warblers**   Very small to sparrow-sized birds with short, slender bills; usually yellow, olive or blue-gray. Most forage rapidly in foliage; some hawk for insects; a few walk on ground or skulk in vegetation; some bob or flick tail; a few fan tail.

• **Waterthrushes**   Ground-dwelling warblers with short, slender bills. Usually walk or run on ground near water, bobbing tail.

• **Waxwings**   Crested birds clad in soft browns and grays, with yellow tips to tail feathers. Travel in flocks and have bounding flight. Feed on berries and at flowers; may hawk for insects.

Woodpeckers   Mainly black-and-white birds with vertical posture and strong, chisel-like bills. Hitch up tree trunks; drum on wood; have bounding flight.

• **Wrens**   Stocky, brown birds with large heads, slightly downcurved bills, and short tails. Often skulk in vegetation; hold tail over back.

A Guide to Birds

Black Tern
Chlidonias niger

Robin-sized
Freshwater Marshes; Inshore
Waters; Salt Marshes; Lakes and
Reservoirs; Open Ocean

Field Marks
9¾″. A small, dark tern that nests in loose colonies in freshwater marshes and moves to coastal or offshore waters during migration. Flight agile and swallowlike, often low over water. Dives for fish, but also captures insects in flight or plucks them from surface.
Breeding adult (1–3) largely black, with gray upper-wing coverts and wing linings; tail shallowly forked, gray; white undertail coverts; leading edge of wing coverts white (1). Adults in winter have gray back and wings, white underparts with prominent dark patch at side of breast, and dark hood extending below eye. Immature similar to winter adult, but upperparts darker and wing coverts barred with brown.
Call a shrill *kik* or *krik*.

Similar Species
Breeding adult unmistakable. Winter adult is only tern with hooded effect.

Range
Breeds from British Columbia, the Prairie Provinces, and New Brunswick south to central California, Nebraska, Great Lakes, and n. New York.

Sooty Tern
Sterna fuscata

Pigeon-sized
Southern Florida: Open Ocean;
Beaches

Field Marks
16″. A medium-sized, dark-backed tern of ocean waters well offshore.
Flies with deep, strong wingbeats; often circles high in the air over
water. Dives for fish like other terns. Nests in colonies on offshore
islands and isolated ocean beaches.
Adults (1–3) have blackish upperparts, cap, and tail; forehead and
outer tail feathers white; underparts and sides of neck white; bill
black. Immature blackish brown except for white lower belly,
undertail coverts, wing linings; whitish flecks on back, wings.
Call a noisy *wide-awake, wide-awake* or *wacky-wack, wacky-wack*.

Similar Species
Bridled Tern, *Sterna anaethetus*, very similar, but has gray back and
wings, white collar around neck, and more buoyant flight; found
offshore in summer from North Carolina southward.

Range
Breeds on Dry Tortugas off s. Florida and rarely on Gulf Coast in
Louisiana and Texas. Occasionally seen in summer far offshore from
North Carolina to Florida and in the Gulf of Mexico.

Black Skimmer
Rynchops niger

Crow-sized
Inshore Waters; Beaches
and Coastal Dunes; Mudflats and
Tidal Shallows; Salt Marshes

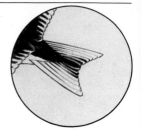

Field Marks
18″. Unmistakable. A large, lanky, long-winged bird of shallow saltwater habitats. Feeds by skimming along surface, plowing long lower mandible through water and snapping up small fish. When standing, has distinctive horizontal posture and short legs. Nests in colonies on protected dunes and beaches. Swift and very graceful in flight.
Adults (1–3) black above, with white underparts and face; bill long and bladelike, red with dark tip, lower mandible longer than upper; legs red, short. Immature similar, but mottled above and often with shorter bill.
Calls are short, barking notes.

Range
Breeds along coasts from extreme s. California and Massachusetts south into tropics. Winters in coastal s. California and from Virginia southward.

Caspian Tern
Sterna caspia

Crow-sized
All freshwater and saltwater
habitats except Open Ocean
and Streams and Brooks

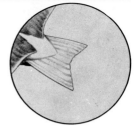

Field Marks
21″. Our largest tern, bulky and nearly the size of a Herring Gull, it
has a stout, spear-shaped bill and a shallowly forked tail. Found in a
variety of freshwater and marine habitats. Nests in small colonies
on sandbars or gravel banks, or in larger colonies of gulls or terns.
Dives for fish like other terns.
Breeding adults (1–3) have gray mantle, black cap with slightly
crested look, and heavy, coral-red bill; extensive blackish on
undersides of primaries (2); tail shallowly forked. Nonbreeding adults
and immatures similar, but cap dusky or streaked; forehead never
wholly white.
Usual call a harsh *kaaar*, deeper than calls of other terns.

Similar Species
Royal Tern smaller, more slender, with little black on underwing,
distinct bushy crest, white forehead during most of year, and orange
bill.

Range
Breeds locally from Washington, Oregon, central Canada, Great
Lakes, and Maritime Provinces south to Mexico, Gulf Coast, and
Florida. Winters from California, Gulf Coast, and North Carolina
southward.

Common Tern
Sterna hirundo

Pigeon-sized
Most freshwater and saltwater
habitats

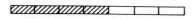

Field Marks
14½″. A medium-sized, slender, swallowlike tern with black cap and
deeply forked tail. Flight swift and graceful. Hovers and dives for
small fish. Nests in colonies on islands, beaches, sandbars, gravel
banks, and occasionally in marshes. Vigorously attacks human
intruders and predators in its colonies.
Breeding adults (1–3) have pale gray back and wings, darkest on
primaries; tail white, cap black; bill slender, red with dark tip; feet
red. Winter adults and immatures similar, but cap reduced to blackish
nape and line through eye; dusky bar on wing coverts; bill and feet
blackish.
Call a harsh, grating *kee-aaarr*; also a sharp *kip, kip*.

Similar Species
Forster's Tern has whitish primaries, gray tail, orange bill with dark
tip in breeding plumage, heavy dark patch through eye and ear
coverts in winter plumage, and different call. See Arctic and Roseate
terns.

Range
Breeds from n. Alberta, Ontario, and Newfoundland south to
Dakotas, Great Lakes (rare), and Gulf of St. Lawrence; also along
Atlantic Coast south to Carolinas. Common migrant along Pacific
Coast. Winters in tropics, rarely on Gulf Coast.

Arctic Tern
Sterna paradisaea

Pigeon-sized
Rocky Shores; Beaches; Inshore
Waters; Open Ocean

Field Marks
15½". A slender, coastal or seagoing tern with black cap, long, deeply forked tail, and very short legs. Hovers and dives for fish. Nests in colonies on rocky or sandy islands, or isolated beaches. Attacks human intruders and predators in its colonies.

Breeding adults (1–3) have pale gray back and wings, with primaries the same color as the rest of wing; tail white; underparts strongly tinged with gray; bill entirely red, slender; feet red. From below, primaries appear translucent in flight. In standing birds, tail extends beyond wing tips, legs appear very short. Winter adults and immatures similar, but underparts white, black cap reduced to dark nape and line through eye, bill black; immatures have whitish secondaries.

Call a rasping *kee-ya*, higher-pitched than Common Tern's.

Similar Species
Common Tern has primaries darker than rest of wing and without translucent effect, longer legs, and a different call; breeding adult has red bill with dark tip and underparts less strongly tinged with gray. Forster's Tern has whitish primaries, gray tail; breeding adult has orange bill with dark tip; winter birds have black ear patch.

Range
Breeds from Alaska and Maritime Provinces south to Massachusetts. Migrates over open ocean. Winters in Southern Hemisphere.

Forster's Tern
Sterna forsteri

Pigeon-sized
Freshwater Marshes; Salt
Marshes; Inshore Waters; Lakes
and Reservoirs

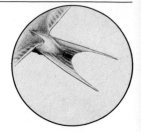

Field Marks
14½". A slender, marsh-dwelling tern with black cap and deeply forked tail. Flight swallowlike and graceful. Dives for fish, but also captures flying insects like other marsh terns. Nests in loose colonies on inland or coastal marshes.
Breeding adults (1–3) have pale gray back and wings, whitish primaries, and gray tail; cap black; bill slender, orange with dark tip; legs orange. Winter adults and immatures similar, but dark cap replaced by heavy blackish ear patch.
Call a harsh, nasal *aaaap* or *aaar*; also a high-pitched *kyer.*

Similar Species
Common Tern has dusky primaries, white tail, and different call; bill red with dark tip in breeding plumage; nape and eyeline black in winter plumage; immatures have dusky bar on inner wing. See Arctic and Roseate terns.

Range
Breeds from e. British Columbia and Manitoba south to s. California, Colorado, and Iowa, and locally on Great Lakes; along Atlantic and Gulf coasts from New York to Texas. Winters from central California and Virginia south to Mexico.

Royal Tern
Sterna maxima

Crow-sized
Inshore Waters; Beaches
and Coastal Dunes; Mudflats
and Tidal Shallows

Field Marks
20″. A large tern of sandy beaches and neighboring salt water. Nests
in large, dense colonies on sandbars, islands, and isolated beaches.
Dives for fish.
Spring adults (1–3) have gray mantle, white underparts, black cap,
and bushy crest; shows blackish on upper side of outer primaries, but
little blackish on underwing; tail moderately forked; bill bright
orange, stout, spear-shaped. Immatures and adults in most of year
have white forehead, black of cap not extending to eye.
Call *kirrup* or *kee-er*, higher-pitched than call of Caspian.

Similar Species
Caspian Tern larger and bulkier, with stouter red bill, more shallowly
forked tail, and extensive black on underside of primaries.

Range
Breeds along Atlantic Coast from Maryland south to central Florida
and along Gulf Coast. Winters in s. California and from Virginia
southward. Wanders north to New England and to central coast of
California.

Least Tern
Sterna antillarum

Robin-sized
Inshore Waters; Beaches; Rivers

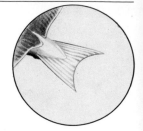

Field Marks
9″. A small tern with rapid wingbeats, usually found near beaches or river sandbars. Neck appears short; when flying over water with bill pointed downward, has distinctive hunchbacked appearance. Feeds by diving and skimming for fish.

Breeding adults (1–3) have gray back and wings, with conspicuous black on outermost primaries; tail short, deeply forked; cap black with white forehead; bill yellow with black tip. Nonbreeding adults similar, but black cap reduced to dark nape and line through eye. Immature like nonbreeding adult, but upperparts mottled.

Calls are a high-pitched *kip-kip* and a piercing *ki-deek!*

Similar Species
All other terns larger, with slower wingbeats and different calls.

Range
Breeds along coast in California and from Maine south to Florida and Texas, and inland along rivers in Midwest and Great Plains. Winters in tropics. Numbers declining in California and inland.

Sandwich Tern
Sterna sandvicensis

*Pigeon-sized
Inshore Waters; Beaches and
Coastal Dunes*

Field Marks
15″. A medium-sized, fast-flying, crested tern of coastal beaches, bays,
inlets, and estuaries. Nests in colonies on sandy beaches, often with
Royal Terns. Dives from air for fish.
Spring adults (1, 2) have gray back and wings with blackish outer
primaries; underparts white; cap black with bushy crest; tail
moderately forked; bill slender, black with yellow tip. Adults during
most of year (3) have white forehead. Immatures similar to
nonbreeding adults, but tail dark rather than white; upperparts
mottled.
Call a grating *kirr-ick*.

Similar Species
No other tern has black bill with yellow tip. Gull-billed Tern has
shorter, stouter, all-black bill and paler upperparts.

Range
Breeds from Virginia south to s. Florida and Texas. Winters on Gulf
Coast and in Florida. Wanders north to New England.

Gull-billed Tern
Sterna nilotica

Pigeon-sized
Inshore Waters; Beaches and
Coastal Dunes; Salt Marshes

Field Marks
14″. A stocky, medium-sized, pale tern with a shallowly forked tail.
Inhabits salt marshes and lagoons, and nearby beaches, bays, and
inlets. Dives for fish, but more often seen hawking for insects. Nests
on beaches in single pairs or in colonies with other terns.
Breeding adults (1–3) have very pale gray back and wings, white
underparts, black cap, and stout all-black bill. Nonbreeding adults and
immatures similar, but cap reduced to smudge through eye and dusky
nape; immatures have mottled upperparts.
Call a rasping *za-za-za* or *kay-deck, kay-deck.*

Similar Species
No other tern has a stout all-black bill. Breeding Roseate Tern has
pale upperparts, but black bill longer and more slender, tail long and
deeply forked. Nonbreeding Forster's Tern has black ear patch,
slender bill, and deeply forked tail.

Range
Breeds at Salton Sea, s. California, and from Long Island (rare) and
New Jersey south to Florida and Texas. Winters from Gulf Coast and
Florida southward. Exclusively coastal except in s. California and
Florida.

Roseate Tern
Sterna dougallii

Pigeon-sized
Beaches; Rocky Shores; Open
Ocean; Inshore Waters

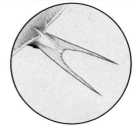

Field Marks
15½″. A medium-sized, very pale, coastal or marine tern with very
long, deeply forked tail. Dives for fish like most other terns. Nests in
small colonies or with Common Terns on beaches and islands.
Uncommon, and seldom seen away from nesting colonies.
Breeding adults (1–3) have very pale gray back and wings, with
outermost few primaries darker; cap black; underparts white, faintly
tinged with pink; bill largely black, long; feet red; tail white, very
long, extends well beyond wing tips in standing birds. Winter adults
and immatures similar, but black cap reduced to dark nape and
hindcrown; immatures have scaly back and black feet.
Call a soft, musical *chivick* or *chewick*.

Similar Species
Common, Forster's, and Arctic terns have darker back and wings,
shorter tails, more red or orange in bill during breeding season, and
different calls.

Range
Breeds in scattered colonies along Atlantic Coast from Nova Scotia
and Maine south to New York. Winters in tropics.

Little Gull
Larus minutus

Robin-sized
Inshore Waters; Lakes and
Reservoirs; Freshwater Marshes

Field Marks
11″. A very small gull, usually seen in winter feeding over tidal channels, at sewage outlets, or over shallow bays, alone or with Bonaparte's Gulls. Nests in freshwater marshes, often in colonies of marsh-nesting terns.

Breeding adults (1) have solid black head (no eye-ring) and slender black bill; back and wings pale gray, with white tips to flight feathers and blackish wing linings. Winter adults (3) similar, but black hood reduced to grayish cap and dark spot behind eye. Immatures (2, 4) have blackish crown and ear spot, gray upperparts with black M pattern across extended wings, white wing linings, and white tail with black band near squared tail tip.

Call a harsh *ka-ka-ka-ka-ka*.

Similar Species
Bonaparte's and Common Black-headed gulls larger, with stouter bills, broken white eye-ring, conspicuous wedge of white in primaries, and pale wing linings; immatures have black trailing edges on wings.

Range
An Old World species that now breeds locally in Great Lakes region. A few winter on Great Lakes and along Atlantic Coast from New England to Virginia.

Bonaparte's Gull
Larus philadelphia

Pigeon-sized
Inshore Waters; Beaches; Lakes
and Reservoirs; Freshwater
Marshes

Field Marks
13½". A small gull of tidal inlets, sewage outlets, beaches, and
northern marshes. Flight buoyant and ternlike. Flutters over water,
plucking small fish from surface; at inland habitats, captures insects in
flight.
Breeding adults (1, 3) have black head with broken white eye-ring and
black bill; back and wings gray, with bold white wedge in primaries;
undersides of wing pale; neck, underparts, and tail white. Winter
adults similar, but head white with small black spot behind eye.
Immatures (2, 4) like winter adults, but with black bill, dark border
on trailing edge of wing, dark borders on white primaries, dark bar on
inner wing, and black band on tail.
Call a soft but rasping *kik* or *keer*.

Similar Species
Adult Common Black-headed Gull larger, with slender red bill and
blackish undersides to primaries; immature Black-headed has pinkish
bill with dark tip.

Range
Breeds from Alaska across central Canada to southern Hudson Bay.
Winters on Pacific Coast from Washington southward, on Atlantic
Coast from New England southward, and in Mississippi Valley.

Common Black-headed Gull
Larus ridibundus

Crow-sized
Inshore Waters; Beaches; Lakes
and Reservoirs

Field Marks
16″. A medium-sized gull that occasionally appears in migration and
during winter at tidal inlets, harbors, sewage outlets, and lakeshores
with the very similar Bonaparte's Gull. Flight agile; plucks food from
water's surface, and captures insects over fields and in plowed
areas.
Breeding adults (1–3) have dark brown head, appearing blackish, with
white eye-ring; bill red, slender; mantle pale gray with bold wedge
of white in primaries, blackish on underside of primaries; neck,
underparts, and tail white. Winter adult (4) similar, but has white
head with small black spot behind eye. Immature like winter adult,
but has pinkish, dark-tipped bill; white primaries with dark edges;
blackish bar on inner wing and dark trailing edge; black band on tail;
and black bill.

Similar Species
Bonaparte's Gull smaller, more ternlike, with shorter black bill; lacks
blackish undersides to primaries (only tips of primaries dark).

Range
An Old World species that has nested in e. Canada and on Cape Cod.
A few winter on Great Lakes, Atlantic and Gulf coasts.

34

Sabine's Gull
Xema sabini

Pigeon-sized
Open Ocean; Inshore Waters

Field Marks
13½". A small, ternlike arctic gull with a shallowly forked tail; occurs in our area mainly as a migrant far offshore. Flies with delicate and buoyant wingbeats, often close to water's surface.
Breeding adult (1–4) has dark head and dark bill with yellow tip; underparts and tail white; mantle gray; outer primaries black; white triangle on rear edge of wing very conspicuous. Winter adult has dark head reduced to mottling on nape. Immatures similar, but with brown mantle and nape, and narrow black tail tip.

Similar Species
Bonaparte's and Common Black-headed gulls have white triangle near wing tip, rather than rear edge of wing. Immature Black-legged Kittiwake has dark M pattern on mantle; lacks bold white triangle on rear edge of wing.

Range
Breeds in Alaska and arctic Canada. Migrates over open ocean; only rarely seen from land. Winters on southern oceans.

Laughing Gull
Larus atricilla

Crow-sized
Inshore Waters; Beaches;
Mudflats; Salt Marshes

Field Marks

16½". A slender, long-winged, black-headed gull of coastal marshes, shores, and waterfronts; rare inland. Feeds on mudflats and beaches; captures flying insects over marsh grass. Nests in colonies in isolated salt marshes.

Breeding adult (1, 2, 4) has black head with broken white eye-ring; bill red; mantle dark gray with black wing tips; neck, underparts, and tail white. Winter adults similar, but black on head reduced to grayish smudge behind eye and on nape; bill black. Immatures (3) are dark brown, with white border on trailing edge of wing, white rump, black bill and tail; older immatures have gray back and wings of adult, but retain dark tail.

Call a piercing laugh: *ha-ha-ha-ha-ha-haaa, haaa, haaa*.

Similar Species

Adult Franklin's Gull has white band separating black wing tip from gray of inner wing; winter birds and immatures have more sharply defined dark hood; immatures have white tail with black band.

Range

Breeds along Atlantic Coast from Maine to Texas; also at Salton Sea in s. California. Winters from North Carolina southward, sparingly farther north.

Franklin's Gull
Larus pipixcan

Crow-sized
Freshwater Marshes; Inshore
Waters; Salt Marshes; Open Areas
and Grasslands

Field Marks
14½". The common black-headed gull of northern grassy marshes and, during migration, on the Great Plains and along the Gulf Coast. Feeds by capturing flying insects; it also follows plows to capture unearthed cutworms and grubs. Pink-bellied spring migrants, flying north in flocks over Plains states, are called "prairie doves."
Breeding adults (1, 2, 4) have black head with a conspicuous broken white eye-ring; bill red, slender; back and wings gray with black wing tip separated from gray by white band; neck, underparts, and tail white; in spring, underparts are tinged with pink. Winter adults similar, but black on head replaced by sharply defined dark hood that extends from crown and nape to eye and ear region. Immatures (3) have distinctly hooded appearance, dark wings with white trailing edges, and black band on white tail.
Call a shrill *ha-ha-ha-ha-ha*, higher-pitched than Laughing Gull's.

Similar Species
Laughing Gull lacks white band separating black wing tip from gray inner wing; winter adults and immatures have less sharply defined blackish hood; immatures have all-black tail.

Range
Breeds from Saskatchewan and Manitoba south to Utah and Iowa. Migrates over Great Plains. Winters chiefly south of our range; sparingly on Gulf Coast.

Lesser Black-backed Gull
Larus fuscus

Crow-sized
Inshore Waters; Beaches; Lakes
and Reservoirs

Field Marks
21″. A rare visitor from northern Europe, found mainly in winter at beaches, harbors, and garbage dumps. Although uncommon, easily located because of some birds' habit of returning to same spot every year.
Breeding adults (2) have white head, underparts, and tail; bill yellow with red spot on lower mandible; eyes yellow-gray, pale; mantle very dark gray or blackish; wing tips black with white spots; legs yellow. Winter adults (1, 4) similar, but with heavy streaking on head and dark smudge around eyes. Immatures (3) mottled brown, with very dark flight feathers in wing; bill blackish, becoming pale at base with dark tip; legs pinkish.

Similar Species
Great Black-backed Gull larger, with stouter bill, darker mantle, and pink legs and feet. Immature Herring Gull similar to immature Lesser Black-backed, but paler on inner part of upper wing.

Range
An Old World species recorded regularly in small numbers on Atlantic Coast and Great Lakes; very rare elsewhere.

Great Black-backed Gull
Larus marinus

Goose-sized
Inshore Waters; Beaches; Rocky
Shores; Open Ocean

Field Marks
30″. Our largest gull, a powerful species found almost exclusively along coast. Flight strong; wingbeats relatively slow. Scavenges like other gulls and steals food from other birds, but also preys on seabirds. Formerly nested south to New England, but range expanding: now nests as far south as Carolinas. Nests in loose colonies, often with Herring Gulls.
Adults (1, 2, 4) have white head, underparts, and tail; wings and tail blackish with white spots at wing tip; bill stout, yellow with red spot on lower mandible; legs and feet pink. Immature (3) mottled with dark brown above; has contrasting whitish rump, dark tail band, much white on head and underparts, and blackish bill.
Call a hollow *skow* or *skowp*, much deeper than call of Herring Gull.

Similar Species
Lesser Black-backed Gull smaller, with somewhat paler mantle, more slender bill, and yellow legs; immature darker.

Range
Breeds along Atlantic Coast from Labrador south to Carolinas; rarely on Great Lakes. Winters on Great Lakes and from Newfoundland south to Carolinas, rarely farther south.

Herring Gull
Larus argentatus

Goose-sized
All freshwater and saltwater
habitats; Urban Areas; Parks

Field Marks
25″. A large, gray-mantled gull, abundant near water in both coastal
and inland areas. Often gathers in wheeling flocks and soars along
cliffs and over buildings. Swims easily; forages at surface and follows
ships for food thrown overboard. Nests in colonies on rocky or sandy
shores and islands.
Breeding adults (1, 2, 4) white, with gray back and wings; wing tips
black with white spots; bill stout, yellow with red spot on lower
mandible; eyes yellow; legs and feet pink. Winter adults similar,
but head streaked with brown. Immatures (3) brown, darkest on
primaries and tail, with gray appearing on back and wings as
adult plumage is acquired; bill pinkish with dark tip; legs and
feet pinkish.
Call a loud, yelping *kyow kyow kyow kyow* or *eeyah eeyah eeyah*.

Similar Species
See Thayer's and Ring-billed gulls.

Range
Breeds from Alaska and arctic Canada south to s. British Columbia,
w. Saskatchewan, Montana, Great Lakes region, and n. New
England, and on Atlantic Coast to North Carolina. Winters from s.
Alaska, Great Lakes, and Labrador southward, especially along
coasts and on open inland waters.

Ring-billed Gull
Larus delawarensis

Crow-sized
All freshwater and saltwater
habitats; Open Country; Urban
Areas; Parks

Field Marks
17½". A medium-sized gull whose buoyant flight enables it to pluck food items from the water's surface more easily than larger gulls. Also forages on foot in plowed fields, parks, and even parking lots. Nests in large colonies on inland lakes.
Breeding adults (4) have white head, neck, underparts, and tail; mantle gray; wing tips black with large white spots; extensive blackish on underside of primaries; bill yellow with narrow black ring; feet yellow. Winter adults (1, 2) similar, but streaked on head. Immatures (3) have pinkish bill with black tip, gray-brown upperparts, whitish underparts, white tail with narrow black band near tip, and pinkish legs.
Call a shrill *kyow kyow kyow* and various squealing notes, all higher-pitched than calls of Herring Gull.

Similar Species
Herring Gull larger; stouter bill lacks black ring; primaries have less black on undersides.

Range
Breeds from Washington, central Saskatchewan, and central Manitoba south to ne. California, Wyoming, and ne. North Dakota; also in Great Lakes region and Maritime Provinces. Winters from sw. British Columbia and Washington, Great Lakes region, and Nova Scotia southward.

Black-legged Kittiwake
Rissa tridactyla

Crow-sized
Open Ocean; Inshore Waters;
Rocky Shores and Coastal Cliffs

Field Marks
17″. An agile, oceanic gull that breeds on northern sea cliffs and ranges widely at sea off our coasts in winter. Flight agile, on quick, shallow wingbeats. Follows ships, snatching food from water's surface, even diving from the air like a tern.
Breeding adults (1, 4) have white head, underparts, and tail; bill yellow, short; mantle gray with sharply defined black wing tips that show no white; legs black. Winter adult similar, but with grayish smudge on back of head and nape. Immature (2, 3) has strong, blackish M pattern across gray mantle, dark spot behind eye, blackish bar across nape, and narrow black tail tip.
Call a nasal *kitti-wake*, often heard in chorus at nesting colonies.

Similar Species
Winter Sabine's Gull has bold white triangle at rear of wing.

Range
Breeds on rocky cliffs from Alaska and arctic Canada south to Gulf of St. Lawrence and Nova Scotia. Winters on ocean south to s. California and Maryland, rarely farther south.

Iceland Gull
Larus glaucoides

Crow-sized
Inshore Waters; Beaches; Lakes
and Reservoirs

Field Marks
22″. A pale northern gull that winters in harbors, inlets, on beaches, and on lakeshores. Feeds by scavenging and often follows ships. Two forms occur in North America, one with no dark pattern in wings, the other with dark gray in wing tips.

Darker adults (1, 4) have white head, underparts, and tail; bill short and slender, yellow with red spot on lower mandible; eyes brown; mantle pale gray and tips of flight feathers darker gray (not black); legs and feet pinkish. Paler adults similar, but eyes yellow, wing tips with no dark markings. Immatures (2, 3) very pale, with buff or sandy mottling on body, inner wing, and tail, and whitish primaries.

Similar Species
Glaucous Gull similar to pale form but larger, with stouter bill. See also Thayer's Gull.

Range
Breeds in eastern Canadian Arctic. Winters on Atlantic coast from Newfoundland to Virginia, and in small numbers on Great Lakes.

Glaucous Gull
Larus hyperboreus

Goose-sized
Inshore Waters; Open Ocean;
Beaches; Coastal Cliffs; Lakes and
Reservoirs

Field Marks
27″. A large, pale arctic gull that winters in our area, where it is found with other gulls at garbage dumps and in harbors; also forages far at sea more than most other large gulls. Flight heavy and powerful.
Adults (1–3) have white head, neck, underparts, and tail; back and wings pale gray with extensive white in primaries; bill large and stout, yellow with red spot on lower mandible. Immatures (4) very pale, often almost entirely white or finely mottled with pale gray-buff; bill pinkish with dark tip.

Similar Species
Pale form of Iceland Gull smaller, with more slender bill and rounder head. Rare albinos of other gulls lack distinctive pinkish, dark-tipped pattern of bill of immature Glaucous Gull.

Range
Breeds in arctic Alaska and Canada. Winters on both coasts south to n. United States and on the Great Lakes.

Thayer's Gull
Larus thayeri

Crow-sized
Inshore Waters; Beaches; Rocky
Shores; Lakes and Reservoirs

Field Marks
23″. A variable, small-billed, gray-mantled gull of the Arctic,
appearing in winter in the United States and southern Canada, where
it is easily confused with Herring and darker Iceland gulls. Habits like
those of Herring Gull.
Typical breeding adults (3) similar to Herring Gull, but have dark
eyes with narrow, fleshy red eye-ring (Herring has yellow eye-ring)
and little or no black on underside of primaries. Some adults paler,
with dark gray rather than black at wing tips; these birds not always
separable from dark form of Iceland Gull, but Iceland Gull usually has
paler mantle and wing tips. Winter adult (1) has streaked head.
Darker immature like immature Herring Gull, but outer primaries
same color as rest of wing, not darker. Paler immatures (2, 4) like
immature Iceland Gull, but have distinct dark band on tail.

Range
Breeds on islands in Canadian Arctic. Winters on Pacific Coast from
British Columbia southward; in small numbers in interior West, on
Great Lakes, and on Atlantic Coast.

Ivory Gull
Pagophila eburnea

Crow-sized
Rocky Shores; Beaches

1

Field Marks
17″. A rare winter visitor from the Arctic, usually found perched on rocks, buildings, or ice near water. Relatively small head, stubby bill, and short legs give this bird a pigeonlike appearance. A scavenger, often seen tearing at carcasses of dead gulls, ducks, or other water birds.
Adults pure white, with dark eyes, black bill with yellow tip, and short black legs. Immatures (1) more commonly seen in our area than adults, similar in shape, but with blackish face patch and black flecking on wings and at tip of tail.

Similar Species
Pale immature Glaucous and Iceland gulls may appear entirely white, but are much larger than Ivory Gull.

Range
Breeds on islands in Canadian Arctic. Winters south to Maritime Provinces, rarely to Great Lakes and Long Island.

Brown Noddy
Anous stolidus

Pigeon-sized
Southern Florida: Open Ocean;
Beaches

Field Marks
15½″. A medium-sized, dark, stocky tern of tropical or subtropical seas, with a short, pointed or wedge-shaped tail. Comes ashore only to nest. Usually seen in flight or perched on driftwood far from land. Seldom dives, but plucks fish from water while in flight.
Adults (1) very dark brown (appearing blackish at a distance) with white cap to level of eye; bill slender; feet black. Immature similar but paler, with white only on forehead.

Similar Species
Black Noddy, *Anous minutus*, rare summer visitor to Dry Tortugas, is smaller (13½″), darker, with more sharply defined white cap and more slender bill.

Range
Breeds on Dry Tortugas off s. Florida. Winters at sea off s. Florida and southward.

Great Skua
Catharacta skua

Crow-sized
Open Ocean

Field Marks
23″. A gull-sized, stocky, dark, broad-winged seabird with a short, hooked bill, found in winter off the Atlantic Coast. Steals food from other seabirds; sometimes seen following ships in pursuit of feeding gulls. In flight, has a distinctly hunched appearance, as do other skuas.
Adults (1) and immatures are dark brown, streaked or mottled with warmer brown, but may look entirely black at a distance; bold white flash at base of primaries.

Similar Species
South Polar Skua, *Catharacta maccormicki* (2, 3), a summer visitor from the Southern Hemisphere, is uniform gray or brown, not streaked. Dark immature gulls lack conspicuous white bases to primaries.

Range
Breeds in northeastern Atlantic. Winters off northern Atlantic Coast south to Maryland.

Pomarine Jaeger
Stercorarius pomarinus

Crow-sized
Open Ocean; Inshore Waters;
Rocky Shores; Beaches

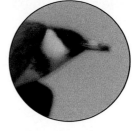

Field Marks
21″. The largest and bulkiest of the three jaegers, found mainly at sea, but also seen from shore, especially along Pacific Coast. Steals food from other seabirds, pursuing them with steady, powerful wingbeats. May follow ships. Adults have long, twisted, spoon-tipped central tail feathers.
Light-phase adults (1–3) have stout bill, black cap, and white cheeks; upperparts brown with flash of white in outer primaries; underparts whitish with dark breast band. Dark-phase birds similar in shape; dark brown with black cap and white flash in primaries. Immatures dark, lack spoon-shaped central tail feathers; best told by shape and behavior.

Similar Species
Parasitic Jaeger smaller, with thinner bill, narrower wing bases, less white in primaries, and sharply pointed central tail feathers; light-phase adults have less pronounced breast band. See Long-tailed Jaeger.

Range
Breeds in Alaska and arctic Canada. Common migrant off both coasts; rare in interior. A few winter on Pacific Coast.

Parasitic Jaeger
Stercorarius parasiticus

Crow-sized
Open Ocean; Inshore Waters;
Rocky Shores; Beaches; Lakes and
Reservoirs

Field Marks
19″. This medium-sized jaeger is the one most often seen from land.
Pursues and steals food from other birds with swift, agile, twisting
flight. Wings narrow, most noticeably at base, and usually angled.
Adults have short, pointed central tail feathers.
Light-phase adults (1, 2) have slender bill, dark cap, and whitish
cheeks; upperparts dark brown with flash of white in primaries;
underparts whitish with broken or smudgy breast band. Dark-phase
adults (3) similar in shape, but wholly dark brown except for black cap
and flash of white in primaries. Immatures dark, lack longer central
tail feathers; best told by shape and flight.

Similar Species
Pomarine Jaeger larger, bulkier, with stouter bill, broader bases to
wings, and spoon-tipped central tail feathers. Long-tailed Jaeger
smaller, more slender, with very long central tail feathers; seldom
seen from land.

Range
Breeds in Alaska and arctic Canada. Common migrant off both coasts;
occasional on Great Lakes. Rare in winter on both coasts.

Long-tailed Jaeger
Stercorarius longicaudus

Pigeon-sized
Open Ocean; Inshore Waters

Field Marks
22″ (including long central tail feathers). A graceful, slender, small-headed jaeger. Flight ternlike. Unlike other jaegers, seldom harries other seabirds for food. Usually seen foraging over ocean; rarely from land. Adults have very long, pointed central tail feathers.
Light-phase adults (1, 2) have small, sharply defined black cap and white collar; upperparts dark gray-brown; underparts pure white or whitish, without breast band. Dark phase extremely rare. Immatures dark gray-brown and lack long tail feathers; best told by slender shape, ternlike flight.

Similar Species
Pomarine and Parasitic jaegers larger, with dark breast band and shorter central tail feathers.

Range
Breeds in Alaska and arctic Canada. Rare migrant off both coasts. Winters in southern oceans.

Magnificent Frigatebird
Fregata magnificens

Goose-sized
Southern Florida: Open Ocean;
Inshore Waters; Beaches

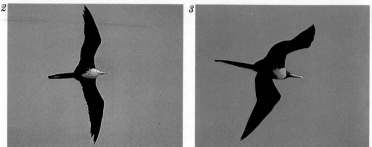

Field Marks
40″. Unmistakable. A large, slender black seabird with long, narrow, pointed wings bent sharply at wrist, long, deeply forked tail, and long, slender bill, hooked at tip. Usually seen soaring or flying with deep, slow wingbeats over ocean or inshore waters, where it obtains food by chasing and robbing other seabirds or dipping down and snatching food from the surface. Perches on bushes in breeding colonies in Florida Keys.
Adult males (1) wholly black, with red throat pouch that is greatly inflated during courtship. Adult females (3) similar, but lack throat pouch, have white breast. Immatures (2) like female, but entire head as well as breast white.

Range
Breeds locally on Marquesas Keys, off s. Florida; seen regularly on both coasts of central and s. Florida, occasionally farther north (rarely to North Carolina); rarely off s. California. Widespread in tropical oceans.

White-tailed Tropicbird
Phaethon lepturus

Crow-sized
Open Ocean

Field Marks
30″. A medium-sized, largely white seabird with very long, streamerlike central tail feathers and long, pointed wings.
Uncommon; usually seen far out at sea. Flight agile, with rapid continuous wingbeats and occasional glides; often turns and twists before diving into water for fish.
Adults (1–3) mainly white, with diagonal black bar on upper surface of wing, black patch at tip of wing, and small black patch through eye; bill orange. Immature similar, but lacks tail streamers, has yellow bill and narrow black barring on upperparts.

Similar Species
Adult Red-billed Tropicbird, *Phaethon aethereus*, very rare visitor off North Carolina, similar but larger (40″), lacks black bar on inner wing, has back barred with black; immature very similar to immature White-tailed, but more finely barred on back.

Range
Uncommon offshore during summer months from Maryland south to Florida, and all year in Gulf of Mexico.

Brown Booby
Sula leucogaster

Goose-sized
Gulf of Mexico: Open Ocean;
Inshore Waters

1

2

Field Marks
30″. A large seabird with pointed wings, a pointed bill, and a body
tapered at both ends. Usually seen far at sea, but occasionally visible
from shore, especially after storms. Plunge-dives for fish, like
Northern Gannet.
Adults (1, 2) dark brown above and on head, neck, and upper breast;
lower breast, belly, and undertail coverts white; shows sharp contrast
between dark upper breast and rest of underparts; wing linings white
with dark borders formed by undersides of flight feathers; bill yellow.
Immature wholly dark, including bill, but lower breast and belly
become paler during first year; feet yellow.

Similar Species
Adult unmistakable. Immature similar to immature Northern Gannet
but smaller, and either wholly dark or showing trace of contrast
between pale underparts and dark upper breast.

Range
Summer visitor to Gulf of Mexico and extreme s. Florida; very rare in
late summer in s. California.

Northern Gannet
Sula bassanus

Very Large
Open Ocean; Rocky Shores and
Coastal Cliffs; Inshore Waters

Field Marks
37″. A large, swift-flying bird of ocean waters, with body tapered at both ends, long, pointed wings, and long, sharply pointed bill. Wingbeats powerful and steady. Dives for fish in spectacular plunges from air. Nests in dense colonies on rocky cliffs and islands. Often seen from shore during migration.

Adult (2) largely white, with black wing tips and golden wash on back of head and neck, which may be hard to see at distance; bill gray. Immature blackish-brown above, paler below, becoming still paler, more mottled as adult plumage is acquired.

Similar Species
Adult Masked Booby, *Sula dactylatra* (1), found offshore in Gulf of Mexico, is similar but smaller (32″), with tail, entire trailing edge of wing, and face blackish; immature has dark head and back, and contrasting white underparts. See Brown Booby.

Range
Breeds in Newfoundland and Gulf of St. Lawrence. Migrates along coast and at sea. Winters from Cape Cod south to Florida and Gulf of Mexico.

Northern Fulmar
Fulmarus glacialis

Crow-sized
Open Ocean

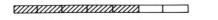

Field Marks
19″. A medium-sized, bull-necked seabird usually seen well offshore, occasionally from shore. Flight stiff-winged, with wings held flat, not bowed as in shearwaters. Sometimes seen in large flocks, especially at fishing vessels. On water, rests with bill pointed downward, head appears rounded.

Typical light-phase adults (1, 3) have white head, neck, underparts, and wing linings; back, wings, rump, and tail gray, with dark wing tips and pale patches in flight feathers. Dark-phase birds (2, 4) variable; usually uniform gray (sometimes very dark), including wing linings, but with pale areas on upper wings. Bill yellow in all phases. Dark phase rare off East Coast.

Similar Species
Gulls more slender, with more leisurely, less stiff-winged flight; lack round-headed appearance. Shearwaters more slender, with narrower wings and thinner bills; wings more bowed in flight.

Range
Breeds on Arctic coasts in Alaska and Canada. Winters south to s. California and mid-Atlantic states.

Wilson's Storm-Petrel
Oceanites oceanicus

Sparrow-sized
Open Ocean; Inshore Waters

Field Marks
7¼". A small blackish seabird that often follows ships, hovering over the surface of the water on quivering wings, with feet touching water. Flight rapid and swallowlike, with continuous wingbeats and only occasional glides. Usually found well offshore, but may enter large bays and harbors.
Adults (2, 4) and immatures blackish, with large white rump patch, rounded wings, and rather long legs that extend beyond squared tail.

Similar Species
Leach's Storm-Petrel, *Oceanodroma leucorhoa* (1, 3), breeds on islands off northeast coast, but less often seen than Wilson's; slightly larger (8"), with shallowly forked tail, more pointed wings, smaller white rump patch, shorter legs, and more erratic flight with deeper wingbeats and much gliding.

Range
Summer visitor from Southern Hemisphere, occurring offshore from Maritime Provinces southward; regularly seen in small numbers off Pacific Coast.

Shearwaters
Puffinus and *Calonectris*

Pigeon- to Crow-sized
Open Ocean

Field Marks
Slender seabirds with long, narrow, pointed wings and thin
bills. Flight rapid, with much gliding and planing close to water's
surface on stiff, bowed wings. Follow fishing boats to feed on scraps.
Usually seen far out at sea, seldom from shore.
Sooty Shearwater, *Puffinus griseus* (6, 8). 19″. Uniformly dark
brown, with whitish wing linings and dark bill.
Cory's Shearwater, *Calonectris diomedea* (2, 4). 18″. Similar to Sooty
but paler above, with light patch at base of tail, white underparts, and
yellowish bill; dark color of crown blends evenly with white throat.
Often "towers" on stiff wings, rising higher above waves than other
shearwaters.
Greater Shearwater, *Puffinus gravis* (1, 3). 19″. Has scaly brown
upperparts; dark brown cap sharply contrasting with white throat and
sides of head; white patch at base of tail; underparts and wing linings
white with dark smudge on belly.
Audubon's Shearwater, *Puffinus lherminieri* (7). 12″. Has shorter
wings than others, more rapid wingbeats; crown and upperparts
blackish brown; underparts, including undertail coverts, white.
Manx Shearwater, *Puffinus puffinus* (5). 15″. Similar to Audubon's
but larger, with longer wings, slower wingbeats, and dark undertail
coverts.

Ranges

Sooty Shearwater breeds in Southern Hemisphere; common off entire Atlantic Coast during warm months; rare in Gulf of Mexico. Also found in Pacific Ocean.

Cory's Shearwater breeds in eastern Atlantic; common offshore from Newfoundland south to Carolinas during summer and fall.

Greater Shearwater breeds in Southern Hemisphere; common off East Coast from Arctic south to Georgia during spring, summer, and early fall.

Audubon's Shearwater breeds in Caribbean; regular far offshore over Gulf Stream from Carolinas south to Florida; very rarely north to Maine and Maritime Provinces.

Manx Shearwater breeds in eastern Atlantic (has been known to breed in Massachusetts and e. Canada); regular offshore from Newfoundland to Carolinas; very rarely to Florida.

Common Murre
Uria aalge

Crow-sized
Open Ocean; Inshore Waters;
Rocky Shores and Coastal Cliffs

Field Marks
17½". A plump, slender-billed, black-and-white seabird most often
seen swimming offshore or nesting in crowded colonies on cliffs. Dives
frequently.
Typical breeding adult (2) has dark brown head, neck, and
upperparts, white underparts, and white border on rear edge of wing;
feet black. "Ringed" form (1) similar, but with narrow white eye-ring
and line behind eye. Winter birds (3) similar to breeding adults, but
throat and cheeks white, with dark line behind eye.
Call a purring *murrr;* various croaks and growls at nesting colonies.

Similar Species
Thick-billed Murre has shorter, stouter bill with white line at base;
face mostly black in winter plumage. Razorbill has much stouter bill.
Black Guillemot smaller, largely black in breeding plumage, more
mottled with white above in winter plumage; feet red.

Range
Breeds along Pacific Coast from Alaska south to central California,
and along Atlantic Coast from Newfoundland to Nova Scotia. Winters
at sea south to s. California and New England.

Thick-billed Murre
Uria lomvia

Crow-sized
Open Ocean; Inshore Waters;
Rocky Shores and Coastal Cliffs

Field Marks
18″. A plump, black-and-white seabird, very similar to Common
Murre; breeding range more northerly, but often appears in winter
farther south than Common Murre. Dives frequently.
Breeding adults (1–3) have dark brown head, neck, and upperparts,
white underparts, and white border on rear edge of wing; bill stout
and pointed, with white line at base; feet black. Winter birds similar,
but throat and lower part of cheeks white.
Calls like those of Common Murre.

Similar Species
Common Murre has longer, thinner bill without white line at base;
sides of face show more white in winter plumage.

Range
Breeds along coast in Alaska and from Canadian Arctic south to Gulf
of St. Lawrence. Winters rarely south to central California, and in
Atlantic to New England and Long Island, occasionally to Maryland,
rarely farther south.

Black Guillemot
Cepphus grylle

Pigeon-sized
Open Ocean; Inshore Waters;
Rocky Shores and Coastal Cliffs

Field Marks

13″. A familiar black-and-white seabird of rocky coasts, usually found closer to shore than other auks. Does not migrate from its northern haunts. Swims with neck held high, and is often seen dipping bill into water. Dives frequently.

Breeding adults (1–3) smooth jet-black overall, with brilliant white upper-wing patch extending to elbow, and white wing linings. Winter adults largely gray and white, with some black. Immatures similar to winter adult but duskier. Feet bright orange-red in all plumages; bill slender, black, and pointed.

Call a high-pitched squealing.

Similar Species

Murres larger, white below in breeding plumage, solid black above in winter plumage. Male White-winged Scoter larger, with white patches on trailing edge of wing, black feet. See Atlantic Puffin.

Range

Breeds along coast from Canadian Arctic south to Maine. Winters at sea off breeding grounds, rarely as far south as Cape Cod.

Dovekie
Alle alle

Robin-sized
Open Ocean; Inshore Waters;
Rocky Shores and Coastal Cliffs

Field Marks
8¼". Unmistakable. A small, chunky black-and-white seabird. Nests in colonies along rocky sea cliffs and on talus slopes; winters far out to sea in North Atlantic. Dives frequently.
Breeding adults (1, 2) have black head, neck, and upperparts; underparts entirely white. Winter birds (3) have white throat, chin, and cheeks, with white forming partial collar around neck. Bill stubby in all plumages.

Similar Species
Murres, Black Guillemot, Razorbill, and Atlantic Puffin much larger.

Range
Breeds in Canadian Arctic. Winters south to New England, rarely to Long Island.

Razorbill
Alca torda

Crow-sized
Open Ocean; Inshore Waters;
Rocky Shores and Coastal Cliffs

Field Marks
17″. An elegant-looking seabird with neat black-and-white plumage and distinctive, arching, white-striped bill. Nests on rocky cliffs; winters farther offshore. Dives frequently; often seen floating with short tail pointed upward.
Breeding adults (1) have black head and upperparts; fine white stripe on rear edge of wings visible at rest; underparts entirely white; bill deep, laterally compressed, black with white band; thin white line runs from top of bill to eye. Winter birds have white throat and ear patch. Immatures similar but with smaller, unmarked bill; lack white stripe from bill to eye.
Calls are low-pitched croaks and growls.

Similar Species
Murres have more slender bills.

Range
Breeds along coast from Canadian Arctic south to central Maine. Winters south to Long Island, rarely off southern states.

Atlantic Puffin
Fratercula arctica

Pigeon-sized
Open Ocean; Inshore Waters;
Rocky Shores and Coastal Cliffs

Field Marks
12½″. Colorful and eye-catching in breeding plumage; famous for its
ability to carry a dozen fish at once in its bright bill. Nests in burrows
along coastal cliffs; in winter, moves to offshore areas near breeding
grounds.
Breeding adults (1) unmistakable, with all-black upperparts, throat,
and crown; face and underparts bright white; bill brilliant red, blue,
and yellow, laterally compressed; feet orange. Winter birds have gray
face and dark bill with inconspicuous yellow tip. Immatures similar to
winter adults but have smaller bill with no yellow.
Call a rasping growl; also low purring notes.

Similar Species
Murres and Razorbill have longer, less stocky bodies. Winter-plumage
Black Guillemot has slender bill, white face, and bold white patches
in wings.

Range
Breeds along coast from Labrador to central Maine. Winters at sea off
breeding grounds, rarely south to Massachusetts and Long Island.

Brown Pelican
Pelecanus occidentalis

Very Large
Inshore Waters; Beaches

Field Marks
42″. A large, heavy-billed bird of coastal regions; rarely found in freshwater areas. Has wingspan of up to seven feet and large, unfeathered throat pouch. In flight, forms loose flocks; soaring alternates with "follow-the-leader" flapping—flock leader flaps first, and others flap in sequence. Often plunge-dives from heights of 20 to 50 feet to capture fish.
Breeding adults (1, 2) large, stocky; silvery brown to gray-brown above and below, with white head; nape and hindneck deep cinnamon-brown; large black throat pouch. Nonbreeding adults lack cinnamon on nape and hindneck. Immature (3) similar but browner overall; lacks white on head and neck.

Similar Species
American White Pelican larger, white with black wing tips; never dives into water from air.

Range
Breeds locally along coast from s. California and North Carolina southward. Wanders north after nesting season to Washington on Pacific Coast, rarely to New England on Atlantic.

American White Pelican
Pelecanus erythrorhynchos

Very Large
Inshore Waters; Lakes and
Reservoirs; Freshwater Marshes

Field Marks
62″. A huge white bird of coastal regions and freshwater marshes.
Like the related Brown Pelican, alternately soars and flaps in flight,
with flock members flapping in "follow-the-leader" sequence. Flies
with head folded back on neck. While swimming, scoops up fishes by
submerging head and neck; does not plunge-dive.
Breeding adults (1, 3) very large, mainly white, with black wing tips,
yellow feet, and unfeathered yellow throat pouch; yellow bill has
raised plate midway along upper mandible. Nonbreeding adults (2)
similar; bill slightly duller, without plate. Immatures duskier with
gray bill.

Similar Species
Whooping Crane has long, slender neck, slender bill, and long black
legs. Wood Stork also soars, but has dark head and neck held out
straight in flight; long dark legs. Snow and Ross' geese much smaller,
with longer necks and stubby pink bills. Brown Pelican may look pale,
but never white with contrasting black wing tips.

Range
Breeds locally from central British Columbia, n. Alberta, and
Minnesota south to n. California, Colorado, and Texas Gulf Coast.
Winters from s. California and along Gulf Coast from Florida and
Texas southward.

Great Cormorant
Phalacrocorax carbo

Very Large
Inshore Waters; Rocky Shores and
Coastal Cliffs

Field Marks
36". A large, dark, long-necked bird of coastal areas. Often seen
perched on rocks or buoys; like all cormorants, spreads its wings to
dry after swimming. The largest North American cormorant, it also
has most northerly winter range, rarely appearing south of coastal
Virginia during cold months.
Adults (2) blackish above and below, with some white on chin and
face, and yellow throat pouch. Breeding birds have white flank
patches. Immatures (1) variable; first-year birds usually brownish-
black above, paler on breast, with white belly; second-year birds more
like adults.

Similar Species
Double-crested Cormorant smaller, more slender, with distinct kink in
neck when in flight, orange-yellow throat pouch; immature usually
palest on neck and breast.

Range
Breeds along coast from Newfoundland south to New Brunswick and
Nova Scotia. Winters from breeding grounds south to New Jersey,
rarely to Virginia, Carolinas, and Georgia.

Double-crested Cormorant
Phalacrocorax auritus

Very Large
All saltwater and freshwater
habitats

1

2

Field Marks
32". A widespread cormorant found in both freshwater and marine habitats, and the only cormorant usually seen in interior wetlands or other inland areas. Easily distinguished from other cormorants in flight by kinked neck. A very good swimmer and diver, it feeds chiefly on fish.
Adults (1) blackish with iridescent greenish or purplish gloss above; throat pouch yellow, bill black. Breeding birds have two inconspicuous tufts above eyes (darkest in eastern birds). Immatures (2) brown above, paler below, with varying amounts of white on neck, breast, and belly, and some yellow on bill. Feet black in all plumages.

Similar Species
Great Cormorant larger and stockier, without kink in neck; yellow throat pouch surrounded by white; birds in breeding plumage have bold white flank patches; immatures usually palest on lower breast and belly. See Olivaceous Cormorant.

Range
Breeds along Pacific Coast from Alaska south to Mexico; from Idaho, Nebraska, and Great Lakes region south to Utah, s. Texas, and Mississippi Valley; along Atlantic Coast from Newfoundland to New York and Chesapeake Bay to Texas. Winters mainly along coast from Alaska to Mexico and from s. New England to Texas.

Olivaceous Cormorant
Phalacrocorax olivaceus

Goose-sized
*Texas and Louisiana: Inshore
Waters; Freshwater Marshes;
Lakes and Reservoirs*

Field Marks
26″. A long-tailed, slender cormorant that is often seen perched on
trees and rocks near water. Widespread in tropics, but reaches our
range in coastal and marshy areas of Louisiana and Texas.
Adults (inset) blackish, with dull yellow throat pouch and base of
lower mandible; breeding birds glossier, with narrow white border
behind throat patch and short white plumes on side of neck.
Immatures (1) dull gray-brown above, somewhat paler below.

Similar Species
Double-crested Cormorant larger and stockier, with shorter tail,
thicker bill and neck; adults lack narrow white border around throat
pouch; flying birds show distinctive kink in neck.

Range
Breeds and winters along coast and on large rivers in s. Texas and
sw. Louisiana; also in s. New Mexico. Wanders throughout the
Southwest.

Anhinga
Anhinga anhinga

Goose-sized
Freshwater Marshes; Inshore
Waters; Southern Wooded
Swamps; Salt Marshes; Lakes

Field Marks
35″. A long-necked, dark water bird, resembling a cormorant at first glance, but easily identified by long tail, slender neck, and unique posture, with head and neck forming an S shape. Inhabits fresh- and saltwater areas; frequently swims with body submerged and only head and neck visible above water. Spears fish with long, straight, pointed bill. Soars to great heights. Often perches in trees with wings outspread.
Adult male (2) glossy greenish-black above and below, with black-and-white streaking on upper wings and upper back; pale, thin white plumes on neck in breeding season. Adult female (1) browner overall, with brown head, neck, throat, and upper breast. Both have long, fan-shaped tail. Immatures similar to adult female, but browner.

Similar Species
Cormorants have much shorter neck, tail, and bill; never soar.

Range
Breeds from central Texas, Tennessee, Alabama, and North Carolina south to Mexico. Winters along Gulf and Atlantic coasts north to South Carolina.

Common Loon
Gavia immer

Very Large
Inshore Waters; Open Ocean;
Lakes and Reservoirs

Field Marks
32″. A large, widespread water bird, generally seen paddling about on
northern lakes during breeding season, and on bays, estuaries, and
open ocean in winter. Well known for its loud laughing call, often
given at night. An excellent swimmer and diver, but awkward ashore;
must patter across water's surface to take off.
Breeding adults (2, 3) marked with bold contrasting pattern of black
and white: head and neck smooth, velvety black, with black-and-white
necklace; upperparts heavily checkered; underparts white. Bill black,
long, daggerlike. Winter birds (1) blackish above, white below, with
pale bill. Immatures similar to winter birds.
Call a loud, yodeling laugh, heard on breeding grounds and during
migration.

Similar Species
Red-throated Loon smaller, with uptilted bill; paler above in winter
plumage. Pacific Loon, *Gavia pacifica*, very rare winter visitor from
Pacific Coast, is smaller, with more slender bill, darker upperparts
contrasting more strongly with white underparts.

Range
Breeds from n. Alaska and n. Canada south to Washington, Great
Lakes region, New York, and New England. Winters along both
coasts and on Great Lakes.

Red-throated Loon
Gavia stellata

Goose-sized
Inshore Waters; Open Ocean;
Lakes and Reservoirs

Field Marks
25″. A large diving bird that swims with its bill tilted slightly upward.
Usually seen alone or in pairs, but forms large groups in food-rich
areas and in migration. In flight, has a somewhat humpbacked
appearance, with neck and head held below line of body.
Breeding adults (1, 3) unmistakable: head and neck silvery gray, with
deep rust-red throat patch and black-and-white pattern on nape;
upperparts rather plain brownish-black, with a little mottling. Bill
black, pointed. Winter birds (2) gray on head and nape, with whitish
throat and underparts, and dark blackish upperparts; bill pale.
Immatures similar to winter birds.
Call a short, wailing yelp; various other calls on breeding grounds.

Similar Species
Common Loon larger, with stouter, horizontal bill; winter-plumage
birds show more contrast between dark upperparts and white
underparts.

Range
Breeds in Alaska and Canadian Arctic. Winters along both coasts
from breeding grounds south to Mexico, and on Great Lakes.

Red-necked Grebe
Podiceps grisegena

Crow-sized
Inshore Waters; Lakes and
Reservoirs; Freshwater Marshes

Field Marks
20″. A large grebe with a long, stout, yellow bill. Breeds on marshy
lakes and ponds in northern part of range; winters along coastal areas,
where it is more easily seen and identified. Feeds on small fishes and
aquatic insects, diving below water's surface. Often quite shy.
Breeding adults (1, 2) have rufous neck, whitish cheeks and throat,
dark brown crown and forehead; back and wings brownish; bill yellow.
Winter birds (3) paler, grayer, lacking rufous on neck; cheek patch
still evident. Immatures resemble winter birds but have little white
on cheeks; bill grayish or pale.

Similar Species
Horned Grebe smaller, with more slender neck, shorter, blackish bill;
head black with yellow ear tufts in breeding plumage, cheek and front
of neck pure white in winter plumage. Winter-plumage loons heavier,
with shorter, thicker neck, more distinct contrast between dark
upperparts and white underparts.

Range
Breeds from Alaska, Mackenzie, and Ontario south to Washington, n.
South Dakota, and s. Minnesota. Winters along Pacific Coast south to
Mexico, and on Atlantic Coast from Gulf of St. Lawrence south to
Long Island, rarely to Florida.

Horned Grebe
Podiceps auritus

Pigeon-sized
Inshore Waters; Freshwater
Marshes; Lakes, Reservoirs,
Ponds, and Rivers

Field Marks
13½". Small, ducklike bird of freshwater marshes, lakes, ponds, and
rivers; found in coastal areas in winter. Swims and dives skillfully;
flight direct, on rapid wingbeats, neck and legs outstretched.
Breeding adults (1, 2) have warm reddish neck, breast, and flanks;
back dark, head nearly black, and throat dark with bright golden ear
tufts; bill dark, rather small. Winter birds (3) have dark crown and
nape; cheeks, throat, and breast white; bill slender and more pale than
in breeding birds. Both plumages show white wing patch in flight.

Similar Species
Eared Grebe, *Podiceps nigricollis*, rare visitor from West, has wispy
golden tufts in ear region and black neck in breeding plumage, dark
crown and sides of head in winter plumage. Other grebes larger or
stockier. See Red-necked Grebe.

Range
Breeds from Alaska and n. Canada south to Washington, Montana,
South Dakota, and Wisconsin. Winters along Pacific Coast from
Alaska to Mexico, and on Atlantic Coast from Nova Scotia to Texas.

Pied-billed Grebe
Podilymbus podiceps

Pigeon-sized
Freshwater Marshes; Ponds;
Inshore Waters; Salt Marshes

Field Marks

13½″. A small, plump bird of marshes and ponds; also found in salt
marshes and inshore waters in winter. Sometimes swims with only
head and neck visible above water's surface. Fairly tame for most of
year, but can be quite shy in breeding season, when its call may be
best indicator of its presence.

Breeding adults (1, 2) brownish overall; stout, pale bill encircled by
dark ring; throat and chin black. Winter birds (3) paler, duller brown,
with whitish chin and throat; bill pale; undertail coverts white.
Juveniles marked with gray and black stripes, but short, stubby bill is
always a good field mark.

Call a hollow *cow-cow-cow-cow, cow, cow, cowp, cowp, cowp*, heard
mainly in breeding season.

Similar Species

Least Grebe, occurring in Texas only, smaller, more slender, and
darker, with short, dark bill; eyes yellow or orange. Other grebes
have slender bills.

Range

Breeds from British Columbia, Alberta, Quebec, and Maritime
Provinces south to Mexico, Gulf Coast, and Florida. Winters in
southern part of breeding range.

Least Grebe
Tachybaptus dominicus

Robin-sized
Texas: Freshwater Marshes;
Ponds

Field Marks
10″. The smallest grebe, found only in marshes and ponds of Texas.
Flies infrequently, and can be very shy, often hiding in thick
vegetation along shorelines. Usually seen singly or in pairs. Feeds
chiefly on aquatic insects and small crustaceans.
Breeding adults (3) dark brownish-gray, with black crown and throat,
and slate-gray head and neck; bill short. Winter birds (1, 2) have gray-
brown head and paler neck. Eyes yellow or orange in all plumages.
Most common call a sharp *peek!*

Similar Species
Pied-billed Grebe larger and browner, with white undertail coverts,
stouter, pale bill. All other grebes much larger.

Range
Breeds and winters from s. Texas south into tropics.

Black Scoter
Melanitta nigra

Crow-sized
Inshore Waters; Lakes

Field Marks
19″. A stocky diving duck of lakes and ponds of the Far North; winters along coastal waters, and may form large flocks in prime feeding areas. Characteristically swims with head erect and bill horizontal or pointed upward.
Adult males (1, 2) black overall, with bright yellow-orange knob at base of black bill; in flight, show silvery flash on undersides of wing feathers. Females (3) brown with pale cheeks and blackish or dark brown head; bill dark.
Male whistles and gives a mournful *cooor-loo*; female croaks.

Similar Species
White-winged Scoter larger, with white patches on trailing edges of wings; female has two pale patches on side of face. Surf Scoter larger, lacks silvery flash on underside of wing feathers; male has white patches on forehead and nape, orange bill; female has two pale patches on side of face.

Range
Breeds locally from Alaska east across Canada to Newfoundland. Winters along Pacific Coast from Alaska south to Mexico, and along Atlantic Coast from Newfoundland south to Carolinas.

Surf Scoter
Melanitta perspicillata

Crow-sized
Inshore Waters; Lakes and
Reservoirs

Field Marks
20″. Sometimes called "skunk-head" for distinctive black-and-white markings on male's head. Most often seen diving for shellfish in breaking waves along coastal areas; both Surf and White-winged scoters use wings to propel themselves underwater.

Adult males (1, 2) black with white patch on forehead and nape; bright orange, black, and white bill with slight knob extends onto sides of face. Females (3) dark brown with darker crown; pale patches on sides of head and at base of large, dark bill. Immatures similar to females, although immature males may be somewhat darker about the face.

Similar Species
White-winged Scoter has white patches on trailing edge of wings. Black Scoter smaller, with silvery undersides to wing feathers; male has conspicuous yellow knob on bill; female has pale cheek contrasting with darker crown.

Range
Breeds in Alaska and Canadian Arctic east to Labrador. Winters along Pacific Coast from Alaska to Mexico, and along Atlantic Coast from Newfoundland to Florida; also along Gulf Coast.

King Eider
Somateria spectabilis

Crow-sized
Rocky Shores; Inshore Waters

Field Marks
22″. A beautiful sea duck of the Far North; small numbers occasionally
wander as far south in winter as New Jersey and the Great Lakes.
Even from a distance, profile of King Eiders unmistakable. Often seen
well out to sea; dives to depths of more than 150 feet to take shellfish
and crustaceans.
Adult males (1, 2) unmistakable; have large, bright orange bill shield,
soft blue-gray head, greenish-white cheeks, deep black back, and
white breast and flank patches. Females (3) mottled brown and black
overall, with steep forehead profile; bill dark; at close range, flank
markings can be seen to be crescent-shaped. Immatures similar to
female; immature male has yellow-orange bill.

Similar Species
Male Common Eider has white back, lacks orange shield on forehead;
female has longer, sloping forehead profile, black bars (not crescents)
on flanks.

Range
Breeds in n. Alaska and Canadian Arctic, south to n. Manitoba.
Winters along coast in Alaska, rarely south to California, and along
Atlantic Coast from Labrador south to Long Island, rarely to
southern states.

Common Eider
Somateria mollissima

Goose-sized
Rocky Shores; Inshore Waters

Field Marks
24". Largest duck in North America. A heavy-bodied sea duck of
rocky shores and inshore waters; flocks of several thousand often seen
in coastal areas, diving for shellfish. Flocks fly single-file along
coastlines, low over the water. On water, usually points bill
downward.
Adult males (1, 2) have bright white back, breast, and cheeks; crown,
flanks, and rear end black; forehead profile long and sloping. Females
(3) also have sloping profile; body rich cinnamon-brown above and
below, with black bars throughout. Immature males have dark brown
body; head somewhat paler, and breast white.
Courting male gives a low moan.

Similar Species
Male King Eider has black back, orange frontal shield; female has
steeper forehead profile, crescent-shaped markings on flanks.

Range
Breeds along coast from Alaska and arctic Canada south to Maine.
Winters from Alaska south to British Columbia (rare), and from
Labrador south to Long Island.

Ring-necked Duck
Aythya collaris

Crow-sized
Lakes and Reservoirs; Inshore
Waters

Field Marks
17″. A small diving duck of lakes, reservoirs, and inshore waters; more at home in forested areas than most other diving ducks. Male's bright white ring around gray bill is best field mark; chestnut collar visible only at close range.

Adult males (1, 2) have black back and breast, pale gray flanks with vertical white blaze near breast; in flight, gray wing stripe visible on black wings. Females (3) dark gray-brown with pale eye-ring. Both sexes have gray bill with white ring and black tip; high, steep forehead; and peaked crown.

Similar Species
Tufted Duck has wispy crest on back of head, lacks white ring on bill; male has white flanks. Female scaups darker, with white face patch and white stripe in wing; lack white eye-ring.

Range
Breeds from Alaska, Manitoba, Quebec, and Newfoundland south to Washington, Minnesota, Great Lakes region, and n. New England; rarely farther south. Winters locally throughout United States.

Tufted Duck
Aythya fuligula

Crow-sized
Lakes and Reservoirs; Inshore
Waters

Field Marks
17″. An infrequent visitor from the Old World to lakes, reservoirs, and
inshore waters. Surest field mark is wispy crest at back of head, but
this may be difficult to see. Often associates with closely related Ring-
necked Ducks.
Adult males (1, 2) have black back, head, neck, and breast; white
flanks; gray bill with dark tip; white wing stripe visible in flight.
Females (3) have dark brown backs, paler flanks, and gray bill with
dark tip; brown head sometimes shows crest at back. Both sexes have
rounded head.

Similar Species
Ring-necked Duck lacks wispy crest, has white ring on bill. Female
scaups have white face patch; Lesser Scaup has shorter white wing
stripe.

Range
Breeds in Old World. Found in small numbers in fall, winter, and
spring from Alaska to s. California, and from Newfoundland to
Maryland.

Barrow's Goldeneye
Bucephala islandica

Crow-sized
Inshore Waters; Lakes,
Reservoirs, and Rivers; Streams

Field Marks
18″. A stocky diving duck with high forehead and triangular head.
Usually seen singly or in pairs in East; more numerous in West.
Wings whistle during flight.
Adult males (1, 2) have glossy purple-black head with crescent-shaped
white spot near base of short bill; back black; breast white; wings
black with large white patches. Females (3) have brown head, grayish
back and wings, and paler gray breast and flanks; bill yellow or orange
with dark tip; white wing patches and collar visible in flight.

Similar Species
Male Common Goldeneye has glossy, dark green head, round white
spot in front of eye, more black on sides; female usually has less
orange on bill. Heads of both sexes rounder than Barrow's.

Range
Breeds from Alaska south through British Columbia to Oregon and
nw. Wyoming, and in n. Quebec and Labrador. Winters along Pacific
Coast from Alaska south to n. California, in Colorado River valley,
and rarely from Gulf of St. Lawrence to Long Island.

Common Goldeneye
Bucephala clangula

Crow-sized
Inshore Waters; Lakes,
Reservoirs, and Rivers

Field Marks
18½". A chunky, round-headed diving duck of bays, estuaries, lakes,
and other quiet bodies of water. Often seen in small flocks; more
numerous in East than similar Barrow's Goldeneye. Flight very rapid;
wings whistle loudly.
Adult males (1, 2) have glossy, dark green head with round white
patch near base of short bill; breast and sides white; back black; white
areas on wings and on back visible in flight. Females (3) have brown
head, gray back and sides; bill dark with small pale patch near tip;
white areas on wings visible in flight.

Similar Species
Male Barrow's Goldeneye has glossy purple head, crescent-shaped
white spot in front of eye, less black on sides; female usually has more
orange on bill. Both sexes have longer, less rounded heads than
Common Goldeneye. Female mergansers more slender, with longer
bodies and ragged crests. Male Bufflehead much smaller, with large
white patch on side of head.

Range
Breeds from Alaska, Northwest Territories, and Labrador south to
Washington, Wyoming, Minnesota, Great Lakes region, and Maine.
Winters along both coasts and on interior lakes and rivers.

Bufflehead
Bucephala albeola

Pigeon-sized
Inshore Waters; Lakes and
Reservoirs

Field Marks
13½″. The smallest duck in North America; often called the
"butterball." Chunky and energetic; seen most often in winter along
sheltered coastal areas. May be quite tame.
Adult males (1, 2) have dark, glossy, greenish-purple head with large
white patch from behind eye to back of head; back black; breast,
flanks, and sides white; white wing patches visible in flight. Females
(3) dark grayish-brown with narrow, elongated white patch on face;
small white area at base of upper wing visible in flight. Bill short
in all plumages.
Male gives squeaky whistle; female quacks softly.

Similar Species
Male Hooded Merganser has thinner bill, black border on white head
patch, rusty flanks. Male Common Goldeneye larger, with small white
spot in front of eye; female lacks white cheek patch. Female Harlequin
Duck has two or three white spots on side of head.

Range
Breeds from Alaska, Manitoba, and central Quebec south to
Washington, n. Colorado, North Dakota, and s. Quebec. Winters
along both coasts and on interior lakes and rivers.

Oldsquaw
Clangula hyemalis

Crow-sized
Inshore Waters; Lakes

Field Marks
22″. Most often seen in winter along protected inshore coastal areas
and lakes. In all plumages, males recognizable by extremely long
central tail feathers. During courtship in spring, flock members fly
abruptly about, then settle back on the water just as suddenly; males
give odd, yodeling call.
Breeding males (1) blackish with white cheeks, flanks, and rear end.
Winter males (2) have white head with dark patches about cheeks;
back marked with bold black and white; wings dark. Breeding females
gray-brown overall, with less conspicuous cheek patches and pale area
on side of neck. Winter females (3) have white head with irregular
dark patches; upperparts brown; flanks pale. Both sexes show dark,
unmarked wings in flight; bill short.
Call of male a repeated, yodeling *ow-owdle-ow, ow-owdle-ow*, heard
most often in spring.

Similar Species
No other bay duck has combination of white underparts and solid
black wings.

Range
Breeds in Alaska and Canadian Arctic. Winters along Pacific Coast
from Alaska south to n. California, and on Atlantic Coast from
Newfoundland south to Carolinas; also on Great Lakes.

Lesser Scaup
Aythya affinis

Crow-sized
Lakes and Reservoirs; Inshore
Waters; Freshwater Marshes

Field Marks
17½". A small duck of prairie ponds and freshwater marshes; in
winter, found along protected coastal areas, often in large flocks;
sometimes occurs with the closely related Greater Scaup. Pale blue
bill distinctive in all plumages; peaked head a good field mark.
Adult males (1, 2) have blackish head with purplish gloss; breast, tail,
and rump black; flanks white or pale gray; back finely marked in gray
and white, appearing gray at a distance; short white wing stripe
visible in flight. Females (3) have dark brown head, breast, and
upperparts; flanks gray-brown; small white patch at base of bill.

Similar Species
Greater Scaup very similar, but has rounder head, longer white wing
stripe; males have greenish gloss on head, whiter flanks. Ring-necked
Duck has white ring around bill, more peaked crown, gray wing
stripe; male has black back, gray flanks, white stripe at side of breast;
female has paler head, white eye-ring, but lacks white face patch.

Range
Breeds from Alaska and Manitoba south to n. California, Nevada,
Colorado, and Great Lakes region. Winters along coast from British
Columbia and s. New England southward, and on interior lakes and
rivers north to Nevada, Colorado, and Great Lakes region.

Greater Scaup
Aythya marila

Crow-sized
Inshore Waters; Lakes and
Reservoirs; Freshwater Marshes

Field Marks
18″. Most frequently seen in huge rafts of thousands of birds, either on
large inland lakes, on bays, or beyond breakers on open ocean. Also
seen in freshwater marshes, but prefers more open areas than
Lesser Scaup.
Adult males (1, 2) have blackish head glossed with green; pale blue
bill; black breast and rump; grayish back; and white flanks. Females
(3) brown overall, darker on back than on flanks, with white patch
encircling face at base of pale blue bill. White wing stripe visible in
flight in all plumages.

Similar Species
Lesser Scaup very similar, but has more peaked crown, shorter white
wing stripe; males have purplish gloss on head, slightly grayer flanks.
Ring-necked Duck has white ring around bill, more peaked crown,
gray wing stripe; male has black back, gray flanks, white stripe at
side of breast; female has paler head, white eye-ring, but lacks white
face patch.

Range
Breeds in Alaska and Canadian Arctic. Winters along Pacific Coast
from Alaska south to s. California, along Atlantic Coast from
Newfoundland south to Georgia, on Great Lakes, and along lower
Colorado River.

Common Merganser
Mergus merganser

Goose-sized
Lakes, Reservoirs, Ponds, and
Rivers; Streams; Inshore Waters

Field Marks
25″. A large, elegant-looking duck, found in a variety of freshwater habitats. When lakes and ponds freeze, Common Mergansers may feed on large rivers, in flocks of 10 to 20 birds all facing upstream and diving for fish. Long, slender, scarlet bill and large size good field marks for all plumages.
Adult males (1, 2) have dark glossy green head, white breast, flanks, and sides; back black; in flight, extensive white visible on back and inner part of upper wings. Females (3) have brown head with ragged crest; white chin and breast; and grayish upperparts; white speculum visible in flight.

Similar Species
Male Red-breasted Merganser has more ragged crest, streaked breast, gray flanks; female lacks sharp contrast between rusty head and white neck.

Range
Breeds from Alaska, Manitoba, and Newfoundland south to n. California, Arizona, New Mexico, South Dakota, Great Lakes region, and n. New Jersey. Winters along coasts and on interior lakes and rivers north to n. California, Great Lakes region, and s. New England.

Red-breasted Merganser
Mergus serrator

Crow-sized
Inshore Waters; Lakes and
Reservoirs

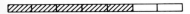

Field Marks
23″. Similar to Common Merganser, but much more likely to be seen
in marine environments than its relative; may occur on freshwater
lakes and reservoirs during migration. Note long, slender build and
tapering red bill in all plumages.
Adult males (1, 2) have dark glossy green head with crest, sometimes
two, at back; neck white; breast brown with blackish streaks; back
black, flanks gray; wings show much white in flight. Females (3)
brownish overall, with ragged crest; small white speculum and white
wing stripe visible in flight. Immatures similar to females.

Similar Species
Male Common Merganser has white breast and flanks, more white in
wings, lacks ragged crest; female has sharp contrast between rusty
head and white neck.

Range
Breeds from Alaska and Canadian Arctic south to n. British
Columbia, Great Lakes, and Maritime Provinces. Winters along coast
from Alaska and Maine south to Mexico.

Northern Shoveler
Anas clypeata

Crow-sized
Freshwater Marshes; Lakes and
Ponds; Salt Marshes

Field Marks
19″. A colorful, medium-sized duck, notable chiefly for remarkable spatulate bill, which is longer than head. Migrates in small flocks of 5 to 10 heavy-looking individuals; feeds by straining small organic matter through the comblike plates within bill.
Males (2, 3) have green head, warm rust-red sides and belly, and white breast. Females (1) mottled warm brownish overall; bill orange around edges. All plumages show green speculum and blue-gray shoulder patch in flight.
Call of female a soft quack; males usually silent except on breeding grounds.

Similar Species
Male unmistakable. Female Blue-winged and Cinnamon teals have smaller, more slender bills.

Range
Breeds from Alaska, Manitoba, and w. Ontario south to n. California, New Mexico, and Missouri; rarely farther east. Winters north to Oregon, Oklahoma, Arkansas, and New Jersey.

Mallard
Anas platyrhynchos

Crow-sized
All freshwater habitats; Mudflats
and Tidal Shallows; Salt Marshes

Field Marks

23″. The largest dabbling duck; a bird of ponds, marshes, streams, city parks, and other freshwater and saltwater habitats. Feeds at surface or by "tipping up" to reach shallow pond bottoms. Like other dabbling ducks, springs directly into air from water. Often seen flying in pairs or trios, as well as in large flocks.

Male (1, 3) has whole head green, with yellow bill, white collar, wholly chestnut breast, gray back and sides, black rump, and pale tail.

Female (2) sandy brown and streaked; bill mottled orange and brown; has very pale, whitish tail. Both sexes have blue speculum with white borders fore and aft.

Usual call of female *quack, quack, quack, quack,* diminishing in both volume and speed; male's calls softer and reedier.

Similar Species

American Black Duck like female Mallard but much darker, with flashing white wing linings, no white borders on speculum. Mottled Duck, of Gulf Coast and Florida, like female Mallard but has unmarked buff throat and cheeks, single white border on rear of speculum.

Range

Breeds from Alaska, Quebec, and central New England south to s. California, Arizona, Texas, central Illinois, and coastal Virginia. Winters north to s. British Columbia, Dakotas, southern Great Lakes region, and New England.

Green-winged Teal
Anas crecca

Crow-sized
Freshwater Marshes; Salt
Marshes; Mudflats and Tidal
Shallows; Lakes and Ponds

Field Marks
15″. The smallest dabbling duck, found in shallow marshes and ponds,
mudflats, and flooded grain fields. Often runs over mudflats in search
of seeds; travels in fast-moving flocks that twist and turn like flocks of
shorebirds.
Male (2, 3) has chestnut head with large green patch around and
behind eye, gray body with vertical white bar at side of pinkish,
speckled breast. Male of Eurasian form, rare winter visitor to Pacific
and Atlantic coasts, has horizontal white line between back and sides.
Female (1) dark gray-brown with buff under tail. Females of
American and Eurasian forms not separable in field. In flight, a small,
dark duck with green speculum, no other pattern on upper wings
or back.
Call of female a soft quack; male gives series of piping whistles.

Similar Species
Other teals have pale blue or whitish patches on upper wings, larger
bills.

Range
Breeds from Alaska, southern Hudson Bay, and Newfoundland south
to n. California, Colorado, Great Lakes, and n. New England.
Winters north to British Columbia, Utah, Texas, and Gulf states, and
on Atlantic Coast to Massachusetts.

American Wigeon
Anas americana

Crow-sized
*Freshwater Marshes; Lakes and
Ponds; Salt Marshes*

Field Marks
19″. Also called the "baldpate," for the male's distinctive white crown.
Feeds chiefly on aquatic plants of lakes, ponds, and both freshwater
and saltwater marshes. Often seen with Redheads, Canvasbacks,
and other diving ducks, but also visits fields to graze.
Males (1, 3) have white forehead and crown and green patch on sides
of head; throat, breast, and sides brownish; wings and back dark
brown or black; white shoulder patch and white belly visible in flight.
Females (2) mottled brownish overall with some gray on head. Both
sexes have bill pale blue with dark tip; white shoulder patch visible in
flight.
Call of male a three-noted whistle with second note higher-pitched;
females give soft quack.

Similar Species
Male Eurasian Wigeon has chestnut head, buff forehead and crown;
females very similar to female Eurasian Wigeon, usually grayer, but
often indistinguishable in field. Gadwall has smaller white patch on
trailing edge of wing, not on shoulder.

Range
Breeds from Alaska, Manitoba, and s. Quebec south to central
California, New Mexico, Minnesota, and Pennsylvania. Winters
mainly along coast from Alaska and Nova Scotia south, and in the
Southwest.

95

Wood Duck
Aix sponsa

Crow-sized
Northern Wooded Swamps;
Southern Wooded Swamps;
Freshwater Marshes; Lakes,
Reservoirs, Ponds, and Rivers

Field Marks
18½". A crested cavity-nesting duck of wooded swamps, marshes, and freshwater areas with plenty of cover. Abundant and familiar; often nests in swamps or near rivers in residential areas. Usually seen in pairs or small flocks, but may form larger flocks in winter.
Males (2, 3) unmistakable, with boldly marked and neatly striped face pattern of green, white, and purplish; throat and upper breast brown, with bold white line separating breast from buff flanks and belly; back and wings dark brown, glossy-looking; bill mainly red. Females (1) dark above, with paler brown breast, pale gray belly; head grayish-brown with distinctive pale patch around eye.
Call of female a piercing *tereeek!* or *woo-eeek!* Male gives a hissing squeal: *ter-eee?*

Similar Species
Male unmistakable. No other freshwater duck has female's bold white eye patch.

Range
Breeds from British Columbia, Saskatchewan, Manitoba, Ontario, and Maritime Provinces south to s. California, Texas, Gulf Coast, and Florida; absent from most of interior West. Winters north to coastal British Columbia, Colorado, s. Iowa, Ohio Valley, and s. New England.

Harlequin Duck
Histrionicus histrionicus

Crow-sized
Rocky Shores; Inshore Waters;
Rivers; Streams

Field Marks
16½". Small, colorful duck, often found in fast-moving streams or in surf along rocky coasts; occasionally found on inland lakes in winter. Swims and walks underwater to forage for aquatic insects.
Males (1, 3) blue-gray, with bold white streaks and patches on face, ear, head, breast, and wings; crown has bright chestnut-red stripe; flanks also bright chestnut; white stripes on back visible in flight.
Females (2) gray-brown overall, with white spots on face under eye and on cheek.
Call a mouselike squeak.

Similar Species
Male unmistakable. Female Bufflehead has single white patch on cheek; in flight shows much white in wings.

Range
Breeds from Alaska south to Oregon and Wyoming, and from eastern Canadian Arctic south to Gulf of St. Lawrence. Winters along Pacific Coast from Alaska south to Mexico, and along Atlantic Coast from Gulf of St. Lawrence south to Georgia.

Hooded Merganser
Lophodytes cucullatus

Crow-sized
Inshore Waters; Lakes and
Reservoirs

Field Marks
18″. A small, handsome, thin-billed diving duck with a fan-shaped
crest. Found only in North America; nests in tree cavities in wooded
areas near lakes and reservoirs, and winters on fresh water and in
coastal lagoons. Usually seen in small flocks.
Adult males (1, 2) have black head with contrasting white crest;
breast white with two broad black lines; back black; belly and flanks
rust-colored; bill dark, thin. Females (3) blackish-brown, darker on
back, with orange or rust-colored crest. Both sexes show white wing
patch in flight.
Courting male gives a froglike croak.

Similar Species
Male Bufflehead has large white head patch without black border,
flashing white flanks, shorter bill.

Range
Breeds from s. Alaska south to Oregon and Idaho, from Manitoba and
Maritime Provinces south to Louisiana and Connecticut, and locally in
interior West. Winters from Alaska south to s. California and from s.
New England south to the Gulf Coast and Florida.

Ruddy Duck
Oxyura jamaicensis

Crow-sized
Freshwater Marshes; Lakes and
Reservoirs; Inshore Waters

Field Marks
15″. A small, heavy-bodied diving duck with a large bill and long tail.
Must patter over water's surface to gain enough momentum for flight.
Nests in marshy ponds and lakes; at other seasons, frequents larger
reservoirs and inshore waters. Often swims with body low in water;
neck appears short in flight.
Breeding males (1, 2) have black head with bold white cheek patch
from blue bill to ear; body deep chestnut; tail long, somewhat darker.
Females (3) have dark grayish-brown cap, pale cheeks with dark line
below eye, and grayish body. Winter male similar to female, but
cheek patch entirely white.

Similar Species
Pied-billed Grebe has stubby, chickenlike bill, lacks long tail feathers
and pale cheeks. See Masked Duck.

Range
Breeds from n. British Columbia, s. Northwest Territories, and w.
Ontario south to s. California, w. Texas, and Iowa; also from Great
Lakes region east to New England. Winters north to s. British
Columbia, Arizona, Texas, Illinois, and s. New England.

Redhead
Aythya americana

Crow-sized
Lakes and Reservoirs; Inshore
Waters; Freshwater Marshes; Salt
Marshes

Field Marks
19″. A common diving duck of freshwater and salt marshes, lakes, and
inshore waters. Feeds mainly on aquatic vegetation, usually feeding
at night; often seen singly or in pairs, except in winter, when flocks of
thousands sometimes band together at feeding grounds.
Males (2, 3) have deep chestnut head, blue bill with black tip; breast
and rear end black; sides and back gray. Females (1) reddish-brown
overall, except for whitish belly; bill blue with black tip. In all
plumages, look for rounded head profile.
In courtship, male gives a catlike *meow*.

Similar Species
Canvasback similar, but with more sloping forehead, long, sloping,
blackish bill; male has whiter back and sides; female has sandy brown
head and neck, grayish back. Female scaups dark, with white face
patch.

Range
Breeds in Alaska and from central British Columbia, s. Northwest
Territories, and Manitoba south to s. California, Colorado, and Iowa;
rarely in Great Lakes region. Winters north to n. California,
Colorado, Missouri, Great Lakes region, and along Atlantic Coast to
s. New England.

Canvasback
Aythya valisineria

Crow-sized
Lakes and Reservoirs; Inshore
Waters; Freshwater Marshes; Salt
Marshes

Field Marks
21″. A large duck with distinctive sloping forehead and long bill. Fast
in flight; often migrates in V-shaped flocks. Most often seen on inland
lakes, freshwater marshes, inland freshwater bays, and other areas
rich in aquatic plant life.
Males (1, 3) have chestnut head and neck; long, sloping black bill;
breast and rear end black; upperparts whitish; white wing linings
visible in flight. Females (2) have paler brown head, neck, and breast;
bill black and long; upperparts pale grayish; wing linings pale.

Similar Species
See Redhead.

Range
Breeds from Alaska, Northwest Territories, and Manitoba south to
n. California, Colorado, nw. Iowa, and s. Minnesota. Winters north to
British Columbia, Arizona, Missouri, Great Lakes, and s. New
England.

Eurasian Wigeon
Anas penelope

Crow-sized
Freshwater Marshes; Lakes and
Ponds; Salt Marshes

Field Marks
20″. A regular visitor to both coasts from the Old World; usually occurs singly in flocks of American Wigeons. Feeds on eelgrass, pondweed, and other aquatic plants, doing most of its grazing at night.
Adult males (1, 2) have rusty head with pale buff-colored stripe on crown and forehead; sides and back delicately patterned in black and white, appearing gray. Females (3) brownish overall, with rusty tinge to head and flanks. Immatures similar to females. All plumages have blue bill with black tip, and white shoulder patch in flight.
Call of male a shrill two-noted whistle; female gives soft quack.

Similar Species
Male American Wigeon has white forehead and crown, broad green stripe behind eye; female very similar to female American Wigeon, but often more reddish.

Range
An Old World species that winters in small numbers along Pacific Coast from Alaska to s. California, and along Atlantic Coast from Maine to Carolinas.

Gadwall
Anas strepera

Crow-sized
Freshwater Marshes; Lakes and
Ponds; Salt Marshes

Field Marks
20″. A common, rather plain dabbling duck of lakes, ponds, and
freshwater and salt marshes. Usually seen in small flocks of 10 to 12
birds. Forages in shallow water, and comes ashore occasionally to
search for grain and seeds. In all plumages, best field mark is white
speculum, unique to the Gadwall.
Adult males (1, 2) grayish overall with some brownish on head;
undertail coverts black. Females (3) mottled brownish overall; narrow
gray bill has orange border.
Female has soft quack; male has various clucks and whistled notes.

Similar Species
No other dabbling duck has white patch in speculum.

Range
Breeds from coastal Alaska, Alberta, Manitoba, and s. Ontario south
to s. California, New Mexico, Missouri, and Great Lakes region; also
along Atlantic Coast from s. New England to North Carolina. Winters
north to se. Alaska, Idaho, Kansas, southern Great Lakes region, and
New England.

Blue-winged Teal
Anas discors

Crow-sized
Freshwater Marshes; Lakes and
Ponds; Salt Marshes

Field Marks
15½". A small brown duck of marshes and ponds throughout most of central North America; especially abundant in prairie potholes and sloughs. Usually travels in fairly small, compact flocks, whose members turn and twist as they fly, revealing pale blue shoulder patches.
Adult males (1, 2) have gray head with white crescent from above eye to just below chin; mottled brownish overall, with small white patch on flanks. Females (3) mottled brown overall; dark line through eye and rather slender bill may be apparent at close range.
Call of male a high-pitched peeping; female gives soft quack.

Similar Species
Female Cinnamon Teal very similar to female Blue-winged, but bill slightly larger, face pattern less distinct. Female Northern Shoveler has much larger, broader bill.

Range
Breeds from Alaska, Manitoba, central Quebec, and Newfoundland south to n. California, New Mexico, Texas, Arkansas, Ohio Valley, and Maryland. Winters mainly along coast from n. California and Virginia southward.

Northern Pintail
Anas acuta

Crow-sized
Freshwater Marshes; Lakes and
Ponds; Salt Marshes

Field Marks
26". A graceful, slender duck with narrow wings; a very fast flyer.
Abundant throughout most of North America, particularly in West in
winter. Inhabits freshwater and salt marshes, lakes, and ponds; feeds
chiefly on vegetation, but will scout along mudflats for small mollusks
and crustaceans in winter.
Adult males (1, 2) have brown head, slender white neck with thin
vertical white stripe extending up to side of head; breast white; body
grayish overall, with some black in wings and fairly long, thin, black
central tail feathers. Females (3) mottled brownish overall, with
slender neck, blue bill, and long tail.
Call of male a mellow, doubled or trebled whistle; female gives a
harsh quack.

Similar Species
Female Mallard stockier, browner, with blue speculum bordered fore
and aft with white. Female Gadwall has white patch in speculum.

Range
Breeds from Alaska and Canadian Arctic south to central California,
New Mexico, Missouri, Great Lakes region, and n. Maine. Winters
north to coastal British Columbia, Utah, Iowa, Ohio, and Long Island.

Black-bellied Whistling-Duck
Dendrocygna autumnalis

Crow-sized
Southern Texas: Freshwater
Marshes; Ponds

1

Field Marks
21″. Unmistakable. A large, long-legged duck with a long neck; often perches in trees, especially on branches overhanging ponds and freshwater marshes. In our range, occurs only in southern Texas and southern Arizona.
Adults (1) are mainly chestnut, with gray on face and throat, and black on lower breast and belly; bill bright red; legs and feet reddish-pink. White on upper wing visible at rest, very conspicuous in flight. Immature has more muted coloration overall, with dark bill, legs, and feet.
Call a shrill whistle, often repeated; most often given in flight.

Similar Species
See Fulvous Whistling-Duck.

Range
A tropical species that breeds in s. Arizona and s. Texas, and winters from s. Texas and Mexico south.

Fulvous Whistling-Duck
Dendrocygna bicolor

Crow-sized
Texas and Louisiana: Freshwater
Marshes; Lakes; Salt Marshes;
Open Areas

Field Marks
20″. Unmistakable. A bright buff-yellow, long-necked duck of
marshes, wet meadows, and rice fields; found only in southern Texas,
Louisiana, and California. Feeds most often at night, frequently on
land. In flight, both head and long legs held below level of body,
creating a distinctive profile.
Adults (1) have bright buff-yellow head, neck, and breast; wings
dark. In flight, show white stripe on flanks and white rump patch.
Immature similar but paler overall.
A squealing whistle: *ka-weeea*, often given in flight.

Similar Species
Immature Black-bellied Whistling-Duck paler, with white wing patch.

Range
Breeds regularly in coastal Texas and Louisiana, and irregularly in
s. California. Wanders widely, rarely north to central California,
Mississippi Valley, and New Jersey.

Cinnamon Teal
Anas cyanoptera

Crow-sized
Freshwater Marshes; Lakes and
Ponds; Salt Marshes

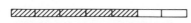

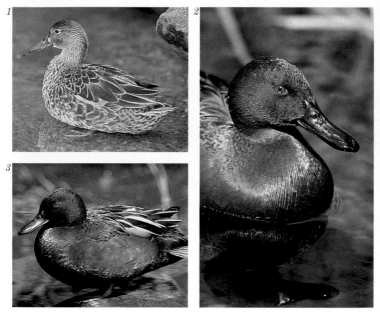

Field Marks
16″. A small duck of marshes and wetlands, especially in arid regions of the continent. Chiefly seen singly or in pairs, or in small family groups in fall; can be quite tame. Feeds by skimming ponds and marshes for sedges, pondweed, and other aquatic vegetation.
Males (2, 3) have deep cinnamon-red head, breast, belly, and flanks; some darker brown on wings; blue-gray shoulder patches visible in flight. Females (1) mottled brown overall. All plumages have long, dark bill, somewhat spoonlike at tip.
Call of male a high-pitched peeping; female gives soft quack.

Similar Species
See Blue-winged Teal. Female Northern Shoveler has much larger, broader bill.

Range
Breeds from s. British Columbia and Saskatchewan south to Mexico and w. Texas, east to western Great Plains. Winters north to n. California, New Mexico, and s. Texas. Casual farther east.

Masked Duck
Oxyura dominica

Pigeon-sized
Texas and Florida: Freshwater
Marshes

Field Marks
13½″. A very small, rather chunky, and somewhat rare duck of tropical wetlands; reaches our area in freshwater marshes of Texas and Florida. A secretive bird, generally hiding in dense marsh vegetation; it can submerge quietly into water like grebes. Rarely seen in flight.
Males (1, 3) mainly ruddy-brown, with black face patch and blue bill with black tip. Females (2) mottled brownish overall, with dark brown or black line through eye and another line crossing cheek just below eye.

Similar Species
Breeding male Ruddy Duck has rufous body, black crown, white cheek patch; winter male has dark cap and white cheeks; female similar to winter male, but has dark stripe on pale cheek. Pied-billed Grebe has stubby, chickenlike bill, lacks long tail feathers and pale cheeks.

Range
A tropical visitor that has bred in coastal Texas and Louisiana; rare in winter in Florida.

Mottled Duck
Anas fulvigula

Crow-sized
Gulf Coast and Florida:
Freshwater Marshes; Lakes and
Ponds; Salt Marshes

Field Marks
22″. A large, mainly brown duck of the Gulf Coast and Florida; found in lakes, ponds, and freshwater marshes, less frequently in salt marshes. From a distance, Mottled Duck may look black. Both sexes resemble female Mallard and female American Black Duck, to which Mottled is closely related.
Adults (1–3) have sandy or buff-brown head; throat boldly mottled in dark brown and lighter brown; greenish speculum bordered at rear with one white bar; bill pale olive-green in male, more yellow in female.
Calls like those of Mallard.

Similar Species
Female Mallard paler, with streaked throat, white borders fore and aft on blue speculum, and orange, mottled bill. American Black Duck has much darker body, purple speculum with no white border, flashing white wing linings.

Range
Breeds and winters along coast of Texas and Louisiana, and in Florida.

110

American Black Duck
Anas rubripes

Crow-sized
Salt Marshes; Mudflats and Tidal
Shallows; Freshwater Marshes;
Lakes and Ponds

Field Marks
23". A close relative of the Mallard, often seen with it on lakes and
ponds, where a few Black Ducks may be present among a huge flock
of Mallards. In such instances, most beginning observers may have
difficulty separating this species from the very similar female Mallard.
In salt marshes and estuaries, however, American Black Ducks are
encountered alone, and are easier to recognize.
Adult male (2, 3) dark blackish-brown overall, with somewhat paler
head and face; wing linings flashing white, speculum purple; bill
yellow; feet and legs red. Female (1) similar with somewhat duller
feet and legs, and dark, mottled bill.
Calls like those of Mallard.

Similar Species
Female Mallard much paler, with white borders on blue speculum.
See Mottled Duck.

Range
Breeds from n. Manitoba, n. Quebec, and Newfoundland south to
Minnesota, Illinois, West Virginia, and North Carolina. Winters from
Kansas, Wisconsin, New York, and New England south to n. Texas,
n. Alabama, and South Carolina.

White-winged Scoter
Melanitta fusca

Crow-sized
Inshore Waters; Lakes and
Reservoirs

Field Marks
21″. A dark brown to black sea duck, common along our coasts in
winter. Breeds on northern lakes and reservoirs; migrates in long,
wavy lines, flying low over water just offshore. Feeds on mollusks and
shellfish, diving to depths of 25 feet.

Adult males (1, 2) have black body and head, with small white mark
behind eye; bill reddish-orange with black knob near base. Females (3)
browner, with darker bill. Immatures resemble females, with more
obvious white patches on face. All plumages show conspicuous white
speculum in flight, often visible at rest.

Similar Species
Surf Scoter similar but lacks white wing patches. Black Scoter
smaller, with silvery undersides to wing feathers; male has
conspicuous yellow knob on bill; female has pale cheek patch
contrasting with darker crown. Breeding-plumage Black Guillemot
smaller, with white patches in wing, not on trailing edge; feet red.

Range
Breeds from Alaska, Northwest Territories, and n. Manitoba south to
ne. Washington, s. Prairie Provinces, and w. Ontario. Winters along
coast from Alaska south to Mexico and from Newfoundland south to
Carolinas.

Greater White-fronted Goose
Anser albifrons

Goose-sized
*Freshwater Marshes; Salt
Marshes; Open Country*

Field Marks
28″. A medium-sized goose of open country, freshwater marshes, and
salt marshes; most common in Midwest and West, but also seen along
southeastern coast in winter. Often feeds in old fields and pastures,
sometimes in flocks of thousands.
Adults (1–3) mainly grayish-brown, with large white patch on face
and pinkish or orange bill; legs and feet orange; belly variably barred
with black; undertail coverts white; white upper-tail coverts visible in
flight. Immatures similar; lack white face, black bars on belly.
Call a very high-pitched bark: *kla-ha*, or *kla-ha-ha*, usually given in
flight.

Similar Species
Immature "Blue" Goose similar, but has dark bill and feet, paler gray
upper-wing coverts.

Range
Breeds in Alaska and Canadian Arctic. Winters from s. British
Columbia south to s. California, in Oregon, Idaho, and Nevada, in
New Mexico, along Gulf Coast of Texas and Louisiana, and rarely on
Atlantic Coast.

Brant
Branta bernicla

Goose-sized
Inshore Waters; Salt Marshes

Field Marks
25". A compact, rather short-necked goose, found almost exclusively on salt water. Flies in compact bunches or loose formations. Feeds in shallow water, especially where eelgrass is common; feeding flocks often noisy.

Adults (1, 2) have dark head and neck with small, indistinct white patch on side of neck; back grayish-brown; belly and flanks paler brown or buff; undertail coverts white. Pacific Coast form, "Black" Brant (3), rare visitor in East, has blackish lower breast and belly, more extensive white collar.

Call a low-pitched, guttural croaking.

Similar Species
Canada Goose usually larger, with longer neck, bold white cheek patch.

Range
Breeds in coastal arctic Alaska and Canada. Winters along coast from Maine to Florida; "Black" Brant from Alaska south to w. Mexico.

114

Canada Goose
Branta canadensis

Goose-sized to Very Large
Freshwater Marshes; Salt
Marshes; Inshore Waters; Lakes
and Reservoirs; Open Country

Field Marks
25–45″. The most familiar goose in North America; easily recognized
even at a distance by the musical honking calls given by V-shaped
flocks. Much regional variation, with several distinct races, differing
chiefly in size. Frequents corn fields and other open-country
environments, as well as marshes and inshore waters, bays, lakes,
and rivers.
Typical adults (1, 2) have black head and neck, with bright,
contrasting white patch on cheeks and chin (sometimes stained
yellow); body gray; undertail coverts white.
Call of larger birds a loud, resonant honking; smaller forms have high-
pitched yelping notes.

Similar Species
Brant smaller and darker; lacks bold white cheek patch; usually found
on salt water. Barnacle Goose, *Branta leucopsis*, rare visitor from Old
World, with whole face white, breast black, belly and flanks white.

Range
Breeds from Alaska and arctic Canada south to California, New
Mexico, Oklahoma, Tennessee, and North Carolina. Winters from
se. Alaska, Utah, Iowa, Pennsylvania, and Maritime Provinces south
to Mexico, Gulf Coast, and Florida.

Snow Goose
Chen caerulescens

Goose-sized
Freshwater Marshes; Salt
Marshes; Open Country

Field Marks
28″. A fairly large, common goose, most often seen in migration, flying in long, wavy lines. Feeds chiefly on rushes and other marsh vegetation, but also searches fields in open country for seeds and grain. Two distinct color phases; dark one, called "Blue" Goose, once considered separate species; rarer in East than white phase. White-phase adults (2, 3) white with black wing tips; bill pink with black "lips"; feet pink. White-phase immature pale sooty gray above, whiter below, with dark bill and feet. "Blue" Goose (1) has dark gray body, white head and neck; belly has variable amount of white; wings gray, with black tips. Immature "Blue" Goose brownish-gray overall, with some white on chin; bill and feet dark.
Call a shrill, yelping *uk-uk*, or *ark-ark*, given in chorus by flocks in flight.

Similar Species
Ross' Goose smaller, with stubbier bill, more rounded head, higher pitched call; scarce in East. Swans much larger and longer-necked; lack black wing tips.

Range
Breeds in Alaska and Canadian Arctic. Winters mainly in California, interior Southwest, along Gulf coast of Texas and Louisiana, and from New Jersey south to Carolinas.

Ross' Goose
Chen rossii

Crow-sized
Freshwater Marshes; Salt
Marshes; Open Country

Field Marks
23″. A Mallard-sized, short-necked white goose, rarely seen east of
Great Plains, but sometimes observed in the lower Mississippi Valley
and along Atlantic Coast. Small flocks of Ross' Geese sometimes
mingle with Snow Geese.
Adults (1, 3) white with black wing tips; bill stubby, pink; neck short.
Immatures (2) pale gray with black wing tips.
Call like a high-pitched bark of a small dog; similar to that of Snow
Goose, but higher and less musical.

Similar Species
Snow Goose larger, with longer bill; lacks rounded head of Ross'
Goose; calls distinctive.

Range
Breeds in Northwest Territories. Winters in isolated areas in
s. Oregon, central California, s. New Mexico, s. Texas, and
sw. Louisiana; winter range expanding, and now occurs rarely
but regularly in East.

Mute Swan
Cygnus olor

Very Large
Inshore Waters; Freshwater
Marshes; Lakes, Reservoirs,
Ponds, and Rivers

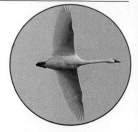

Field Marks
60″. A very large, long-necked swan of freshwater or brackish areas;
easily recognized by its "sailing" posture on water. Introduced from
Europe in nineteenth century; now commonly seen in ponds of public
parks and gardens as well as in the wild. Can be aggressive, especially
in defense of nest.
Adult males (2, 3) white overall, with very long neck; bill pinkish or
orange, with black knob at base. Females similar but smaller, with
smaller bill knob. Immatures (1) gray-brown; darker above, paler
below, with dark bill. Neck shows graceful curve in all plumages.
Makes hissing and grunting notes, but usually silent; wings make loud
rushing (not whistling) sound in flight.

Similar Species
Tundra Swan smaller, with neck usually held straight up, bill largely
black and without knob; usually travels in large and often noisy flocks.
Snow Goose much smaller, with shorter neck, short, pink bill, and
black wing tips.

Range
An Old World species introduced and breeding along Atlantic Coast
from Massachusetts to Virginia and in Great Lakes region.

Tundra Swan
Cygnus columbianus

Very Large
Inshore Waters; Freshwater
Marshes; Lakes and Reservoirs;
Salt Marshes

Field Marks
52″. A melodious, medium-sized swan with a long neck and white
plumage; most widespread swan in North America. Nests on tundra
of Far North; migrates in large flocks, traveling tremendous distances
to winter on marshy lakes and coastal bays.
Adults (1, 2) white with long, rather straight neck; bill black with
small yellow spot in front of eye. Immatures (3) have very pale gray
plumage and pink bill with black tip.
Mellow, high-pitched *hoo-hoo, hoo-hoo-hoo*, often given in chorus by
flocks in flight.

Similar Species
Mute Swan larger, with neck usually arched; bill orange with black
knob; seldom calls or travels in large flocks. Snow Goose much
smaller, with shorter neck, short, pink bill, and black wing tips.

Range
Breeds in Alaska and Canadian Arctic. Winters along Pacific Coast
from se. Alaska to s. California, and in Nevada, s. New Mexico, and
w. Texas; along Atlantic Coast from New Jersey to North Carolina.

Purple Gallinule
Porphyrula martinica

Pigeon-sized
Freshwater Marshes; Salt
Marshes

Field Marks
13″. A heavy-bodied bird of salt marshes, freshwater marshes, and other wetland areas with plenty of vegetation. Somewhat shy; can be difficult to observe. Its very long toes enable it to walk on floating vegetation. Pumps head while swimming.
Adults (1, 2) have bluish-purple head, throat, and upper breast; bill red with yellow tip and pale blue shield at base; back and wings glossy green; legs and feet yellow. Immatures (3) coppery green above and below, with dull brownish or buff head and dark bill; legs and feet yellow.

Similar Species
Common Moorhen brown above with white stripe on sides; head and neck dark gray to blackish. American Coot grayish-black with white bill.

Range
Breeds from s. Texas to w. Tennessee, along Gulf Coast to Florida, and north on Atlantic Coast to North Carolina. Winters on Gulf Coast from Texas to Florida and on Atlantic Coast of Florida. Rarely strays north to Great Lakes region and Canada.

Common Moorhen
Gallinula chloropus

Crow-sized
Freshwater Marshes; Lakes and
Ponds; Mudflats and Tidal
Shallows

Field Marks
14″. A chickenlike, semiaquatic bird of freshwater marshes and
vegetated shores. Lacks webbing or extensive lobes on toes, yet
swims frequently, pumping its head. Retiring, but often comes out
into open. Flight fluttery and awkward.
Breeding adults (1, 2) have head and neck uniform dark gray,
blending into medium gray body; white stripe along top of flanks; back
brown; undertail coverts white and black; some white on belly; yellow-
tipped bright red bill, with broad red frontal shield on forehead; legs
and large feet yellow-green. Winter adults (3) similar, but with very
dusky bill. Juvenile brown above, dull buffy gray with white smudges
below; throat white; face buff; bill dull, with small frontal shield.
Downy chick black with white on throat; bill red with yellow tip.
Calls varied, including raucous squawks, squeaks, and clucking.

Similar Species
American Coot larger with white bill and frontal shield, lobed toes;
lacks white flank stripe. Purple Gallinule has green back; lacks white
flank stripe; solid white undertail coverts; adult blue-purple.

Range
Breeds from California to New Mexico; in the East, from Minnesota to
Vermont, and New Brunswick south to the Gulf Coast. Winters in
California, Arizona, and Texas; also on the Gulf Coast, and Atlantic
Coast to Virginia, occasionally farther north.

121

American Coot
Fulica americana

Crow-sized
Lakes and Ponds; Freshwater
Marshes; Salt Marshes; Inshore
Waters

Field Marks
15″. An abundant swimming bird on fresh water, also winters on salt water. Usually oblivious to humans, but occasionally panics. Plump, odd-looking, with oversized yellow-green feet with lobed toes. Clumsy flight after a running takeoff from water. Feeds at surface or by diving. Pumps head while swimming.
Adults (1, 2) slate-gray with black head and neck; white and black undertail coverts; white bill extends up over forehead as white frontal shield, normally topped by dusky red "callus"; dusky red spot near tip of both mandibles; narrow white stripe along trailing edge of wing shows only in flight. Juvenile (3) smaller and paler gray; initially has much whitish below and on head; frontal shield and white bill color develop gradually. Downy chick blackish with coarse orange-red down on head and neck, yellow down on back; bill red with white tip.
Calls varied, mostly raucous, grating croaks and cackles, repetitious squawks, hoarse squeals, and soft clucking. Vocal at all times.

Similar Species
Generally unmistakable. See Common Moorhen and Purple Gallinule.

Range
Breeds across Canada and south to California and Florida. Winters from British Columbia to California, east to Massachusetts and south to Gulf Coast and Florida. Absent inland from much of Northeast.

Northern Jacana
Jacana spinosa

Robin-sized
Southern Texas: Freshwater
Marshes; Ponds

Field Marks
9½". A small, nervous, rail-like bird; an occasional visitor and rare
breeding bird on ponds and marshes in southern Texas. Has slender
legs and extremely long toes; most often seen walking or running on
lily pads and other floating marsh vegetation. Frequently raises wings
on alighting, displaying bright yellow patch in flight feathers. Usually
found alone, but a pond with breeding birds may have several birds—
a female and more than one male. Flight weak and rail-like; yellow
wing patches and long, trailing legs conspicuous in flight.
Adults (1, 2) chestnut-brown with black head, neck, and breast; bill
and shield on forehead bright yellow; show large yellow wing patches
in flight; legs and feet greenish. Immatures (3) brown above, with
bold white eyebrow and white underparts, dark line through eye;
immatures also have bright yellow wing patches.
Call a grating *week-week-week-week*.

Similar Species
Rails have shorter toes, longer bills.

Range
Rare visitor and breeding bird in s. Texas; also in West Indies,
Mexico, and Central America.

Black Rail
Laterallus jamaicensis

Sparrow-sized
Salt Marshes; Freshwater
Marshes

Field Marks
6″. An extremely secretive, tiny, very compact rail. Runs mouselike
through marsh vegetation. Inhabits coastal and inland marshes, from
salt to fresh. Seldom seen except when flooded out of salt marshes by
very high winter tides. Flies weakly.
Adults (1) blackish above and dark gray below; upper back dusky
chestnut, behind which entire upperparts are speckled with white;
flanks finely barred with white; bill small, black; eyes red; virtually no
tail or neck; legs short. Downy chick minuscule and solid black.
Breeding season calls are a *ki-kee-doo* (male), with middle note
highest and accented; *hoo-hoo-ooo* (female). Also a low *gurrrr*. Most
vocal at night.

Similar Species
Downy chicks of other rails are very small and black, but lack white
speckling and barring and chestnut upper back. Songbirds such as
sparrows and wrens have obvious tails (unless missing), and all have
different colors.

Range
Breeds along Pacific Coast from California south, Atlantic Coast from
Connecticut south; also inland along southern Colorado River and in
Mississippi Valley. Winters in southern part of range to Central and
South America.

Yellow Rail
Coturnicops noveboracensis

Sparrow-sized
Freshwater Marshes

Field Marks
7¼″. One of the most difficult birds to see in North America. A very small, compact rail. Runs like a mouse through grassy swales and similar shallow-water marshy habitats. Winters in wide variety of marshes and fields. Flies weakly, showing white patch in wings. Adult (1) has entire upperparts boldly streaked yellow and blackish, with very fine white perpendicular bars; breast, neck, and face golden-buff, with indistinct bars; paler on throat; belly whitish; bill very short; virtually no tail or neck; brown wings have conspicuous white secondaries in flight. Juvenile duskier, with fine barring from breast to face. Downy chick tiny, solid black.
Typical call on breeding grounds is *tik-tik, tik-tik-tik, tik-tik, tik-tik-tik . . . ,* often given in long series, especially at night; sounds like pebbles being tapped together. Other calls quite varied.

Similar Species
Immature Sora larger; lacks white secondaries; upperparts not boldly streaked yellow; breast dull grayish buff. Songbirds have obvious tails, generally lack white wing patches.

Range
Breeds across central and e. Canada south to North Dakota and the Great Lakes region. Winters rarely along Pacific Coast from central California, and along Atlantic and Gulf coasts from North Carolina south.

125

Sora
Porzana carolina

Robin-sized
Freshwater Marshes; Salt
Marshes

Field Marks
8¾". A medium-sized rail with a short, chickenlike bill. Found in all
types of marshes, but prefers fresh water. Usually hides in marsh
vegetation, but sometimes ventures into open. Flies weakly but
swims well. Short tail often held cocked or flicked forward.
Adults (1) olive-brown above, with broad black streaks and fine white
lines; underparts barred black, white, and gray, strongest on flanks;
cheeks, sides of neck, and breast gray; black face and throat set off
yellow bill. Juvenile similar to adult above, but has buff breast, neck,
and face; whitish throat; bars only on flanks; bill duller. Downy chick
tiny and solid black.
Rapid descending whinny of very short, shrill *dee* notes; plaintive,
ascending *ker-whee*; abrupt *keek!* Other squeaks and whistles.

Similar Species
Yellow Rail very small; white secondaries show in flight; entire
upperparts boldly streaked yellow; golden-buff breast. Virginia Rail
has much longer, drooped bill; longer neck; rufous breast. Common
Moorhen, American Coot, and Purple Gallinule much larger; not
barred below or streaked above; have frontal shield.

Range
Breeds across Canada south to central California and east to
Maryland. Winters from California to Texas; also along Atlantic and
Gulf coasts north to Virginia.

126

Virginia Rail
Rallus limicola

Robin-sized
Freshwater Marshes; Salt
Marshes

Field Marks
9½″. A medium-sized, secretive, and colorful rail; breeds locally in freshwater and brackish marshes, but more widespread in winter, when also found in salt marshes. Flies weakly; often holds short tail cocked or flicked forward. Has a long bill, which it carries angled downward.

Adults (1) have rufous wings, breast, and neck; cheeks gray; chin whitish-buff; olive-brown streaked with black above; flanks barred black and white; fairly long, slightly drooped bill is partly red-orange. Juvenile similar, with underparts variably mottled in black, white, and brown; bill dark. Downy chick tiny and solid black.

Calls varied, including various *kik* notes, grunts, and squeaks; a rapid descending series of *cheg* notes; and *kidik, kidik, . . .* call.

Similar Species
Clapper Rail and King Rail much larger; paler back; cheek and neck about the same color. Sora, Yellow Rail, and Black Rail have bill shorter than head, short neck; wings not rufous. Shorebirds seldom in dense marsh, often gregarious, with different shapes and postures.

Range
Breeds locally across Canada south to California and North Carolina. Winters along Pacific Coast, Atlantic and Gulf coasts from North Carolina.

Clapper Rail
Rallus longirostris

Crow-sized
Salt Marshes; Mudflats

Field Marks
14½″. A large gray-brown rail of coastal salt marshes. Secretive and normally hidden in marsh vegetation; occasionally seen venturing out on mudflats, or when flooded out of marsh by very high tides. Flies weakly; short tail often cocked. Holds dull orange bill angled downward.

Adults (1) brown or gray-brown above, with streaked back; neck and cheeks grayish; throat whitish; breast gray-buff to pale rufous; flanks olive-gray, strongly barred with white; wings brown with olive-brown shoulders; bill fairly long and stout, slightly drooped; sturdy legs and large feet, without webs. Juvenile smaller, with shorter bill; fairly uniform dusky gray-brown; large feet and legs. Downy chick small and glossy black.

Calls varied; typical is decelerating series of 10–25 harsh *kek* notes.

Similar Species
King Rail usually in freshwater or brackish marshes; more rufous; shoulders, neck, and cheeks rufous; usually larger. Virginia Rail smaller; gray cheek contrasts with rufous neck and breast. Other rails much smaller, with short bills. See also shorebirds.

Range
Locally from central California south, and inland at Salton Sea. In East, from Massachusetts to Florida, and along Gulf Coast.

King Rail
Rallus elegans

Crow-sized
Freshwater Marshes; Salt
Marshes

Field Marks
15″. A large, long-billed bird of freshwater and salt marshes. Feeds on small crustaceans, fishes, frogs, and insects, as well as grains and berries. Secretive, but may be heard calling at twilight and dawn. Adults (1) cinnamon-brown or rufous overall, with some dark barring on back and wings; shoulders and breast bright rust without barring. Immatures have somewhat darker backs, paler underparts, without rich cinnamon tones. Downy young glossy black.
Call a chattering, staccato *keh-keh-keh* or *heh-heh-heh.*

Similar Species
Virginia Rail smaller, much more grayish; never shows rich cinnamon tones. Clapper Rail, found only in salt marshes, always has pale olive-brown shoulders; most Clappers much paler than King Rail, but birds in Gulf Coast region may be more reddish.

Range
Breeds from North Dakota to w. New York, and from Massachusetts south to Gulf Coast and Florida. Winters along Atlantic Coast from Connecticut to Florida. Also in Gulf Coast states and in the Mississippi Valley north to s. Illinois.

American Bittern
Botaurus lentiginosus

Goose-sized
Freshwater Marshes; Salt
Marshes

Field Marks

28″. A secretive, marsh-dwelling heron; relatively short-legged and heavy-bodied. Pot-bellied in flight, with pointed-looking head, feet, and wing tips; shows dark primaries. When alarmed, either flushes or freezes with bill and neck pointing upward, blending into reeds. Walks with bill angled upward.

Adult (1) heavily streaked below with rufous-brown on yellow-buff; warm dark brown above, finely marbled on back and wings; broad black stripe on side of yellow-brown neck; sides of chin white; bill yellowish; legs rather short, green. Juvenile similar.

Breeding call an eerie, hollow, low, chanting *oonk-a-chunk*, repeated several times; sounds like an old water pump. Usually heard at twilight or at night. Flushing call a low *kok-kok-kok*.

Similar Species

Immature night-herons grayer brown; lack black neck stripe and contrasting blackish primaries; less pointed-looking; commonly perch in trees. Green-backed Heron much smaller; lacks black neck stripes; has dark green wings and tail. See Least Bittern.

Range

Breeds from se. Alaska and British Columbia east to Newfoundland and south locally to s. California, Texas, and Florida. Winters from British Columbia south to s. California, s. Utah, and New Mexico; along Atlantic and Gulf coasts from s. New England.

Least Bittern
Ixobrychus exilis

Pigeon-sized
Freshwater Marshes; Salt
Marshes

Field Marks
13″. A very small, secretive, marsh-dwelling heron. Light and thin-bodied; often climbs in reeds. Flushes short distances, with quick wingbeats, legs dangling, and neck outstretched. Typically in freshwater marshes, but also visits salt water.
Adult male has black crown and back; chestnut hindneck; yellow-buff below, with white streaking down middle from chin to vent; most of upper-wing coverts yellow-buff, forming a large, chestnut-bordered wing patch, visible at rest or in flight; primaries and secondaries blackish-brown; slender yellow bill and legs. Adult female (1) similar, but with chestnut-brown back and blackish-brown crown; buff colors are duskier; fine streaking below. Juvenile similar to adult female, but throat and breast have broader brown streaks and crown is chestnut-brown,
Gives fast series of low, throaty *coos*; also cackles and *tick* notes.

Similar Species
Green-backed Heron and American Bittern much larger; lack buff wing patches. Rails lack wing patches; different shape.

Range
Breeds locally from Oregon across Canada, south to s. California, Texas, and Florida. Winters along Pacific Coast from n. California and n. Arizona south to Mexico; in the East, in Florida and along the Gulf Coast.

Black-crowned Night-Heron
Nycticorax nycticorax

Goose-sized
All freshwater habitats; Mudflats
and Tidal Shallows; Salt Marshes

Field Marks
25″. A heavy-bodied heron with relatively short, thick legs and bill.
Neck usually well hidden in body contour. Roosts in groups in trees
during day and usually forages at night, using large red eyes to see
prey; sometimes feeds by day. Occurs in freshwater habitats, as well
as mudflats and salt marshes. Flies with deep, steady wingbeats.
Adults (1) has glossy black crown and back; rest of upperparts, wings,
and tail medium gray, blending into whitish underparts; a few thin
white plumes hang from nape; white over base of robust, slightly
downcurved black bill; rather short yellow legs and feet (red during
courtship). Immature (2) dark brown above, streaked with buff and
white; heavily streaked brown on white below; legs and most of bill
yellowish.
Call a distinctive *quock*, often heard at night.

Similar Species
Juvenile Yellow-crowned Night-Heron has small, extensive buff-white
spots on slate-brown back and wings; longer legs and neck; black bill.
See American Bittern.

Range
Breeds throughout much of s. Canada and the United States. Winters
along coasts from Oregon and Massachusetts south to South America.

Green-backed Heron
Butorides striatus

Crow-sized
All freshwater habitats; Mudflats
and Tidal Shallows; Salt Marshes

Field Marks
18″. A small, short-legged heron; compact, but with a deceptively long neck. Found in all freshwater and many saltwater habitats. Generally solitary. Often flushed from vegetation along stream banks. When alarmed, stretches neck, raises crest, and flicks tail downward. Adult (2) has glossy, dark green crown and upper wings; gray-green back; rufous cheeks blend into purple sides of breast; chin white; foreneck and breast streaked; smudgy gray belly; short legs yellow-green to orange; bill mostly blackish. Immature (1) bronze-green above with rufous, buff, and whitish feather edges; buff below, heavily streaked brown from cheeks through breast; yellowish bill and legs. Flushing call is a startling *skeow!*

Similar Species
Least Bittern smaller, with large buff wing patches. American Bittern larger with blackish primaries and black stripe down side of neck; warmer brown. Immature night-herons much larger and heavier; thicker bill. Little Blue Heron larger; longer legs; lacks white stripes below.

Range
Breeds from British Columbia south to s. California and New Mexico; also from North Dakota to New Brunswick, south to Texas and Florida. Winters from central California south on coast and interior; along Atlantic and Gulf coasts from North Carolina to Mexico.

Snowy Egret
Egretta thula

Goose-sized
Freshwater Marshes; Mudflats
and Tidal Shallows; Salt Marshes;
Northern Wooded Swamps;
Southern Wooded Swamps

Field Marks
24″. A gregarious, medium-sized white heron of most fresh- and
saltwater habitats. Often forages actively in shallow water. Flies
with neck coiled tightly and long legs trailing behind; has quicker
wingbeats than Great Egret.
Plumage always entirely white. Breeding adult (1) has long, thin,
filamentous plumes down back ending and curving outwards near tail;
shorter, straighter plumes on nape and over breast; slender,
daggerlike black bill; legs black; feet and lores bright yellow (orange
in courtship). Winter adult has shorter plumes. Immature (2) lacks
plumes; has yellow stripe up rear of black legs; feet yellow.

Similar Species
Immature Little Blue Heron has blue-gray bill with black tip, and dull
greenish legs and feet; less slender. Great Egret much taller, with
black feet and yellow bill; usually stands quietly. Cattle Egret has
yellow to red-orange bill; more compact; shorter, thicker bill and head;
and shorter legs. White-phase Reddish Egret has pink-and-black bill;
heavier. See "Great White Heron."

Range
Breeds locally from central California through to e. Colorado; in the
East, in the Mississippi Valley, and from Long Island south to
Florida. May wander farther north. Winters in California, along
southern Colorado River, and Atlantic and Gulf coasts from Virginia.

Great Egret
Casmerodius albus

Very Large
Freshwater Marshes; Mudflats
and Tidal Shallows; Salt Marshes;
Northern Wooded Swamps;
Southern Wooded Swamps

Field Marks
39″. A white heron almost as tall as Great Blue and "Great White" herons, but much more slender. Normally stands motionless to await prey. Flies with neck in large drooping "sink trap," with long legs trailing; wingbeats slow. In most fresh- and saltwater habitats. Plumage always all-white. Breeding adult (1) has long filamentous plumes down back, extending well past tail; bill yellow to orange, daggerlike; legs and feet black. Winter adults (2) have much shorter and fewer back plumes, usually not apparent. Juvenile lacks plumes; has dark tip on yellow bill.
Usually silent away from colony, but croaks.

Similar Species
"Great White Heron" larger and much heavier, with massive bill and pale yellow legs; Florida only. White-phase Reddish Egret smaller; has pink bill with black tip; "dances" while foraging. Cattle Egret smaller and more compact, with shorter bill and legs, and thick head. Juvenile Little Blue Heron smaller; has blue-gray bill with dark tip and greenish legs. See Snowy Egret.

Range
Breeds from s. Oregon south; from s. Minnesota and Wisconsin south to Texas; and from Long Island south to Florida. Winters along coasts from n. California and New Jersey south; also along Gulf Coast.

135

Cattle Egret
Bubulcus ibis

Crow-sized
Freshwater Marshes; Salt
Marshes; Open Areas and
Grasslands

Field Marks
20″. A gregarious, medium-sized white heron. Usually forages away
from water, especially in grassy areas, unlike most other herons.
Follows and even stands on cattle to catch flushed insects (hence its
name). Also uses wide variety of freshwater habitats, with colonies
placed over water. Compactly built, with shorter, thicker legs and bill
than other egrets. Looks heavy-jowled, and often stands in hunched
posture. Expanding its range.
Breeding adult (1) white, with short filamentous orange-buff plumes
on crown and breast, and longer ones on lower back; bill and legs
yellow (to red-orange during courtship), with duskier feet. Winter
adult (2) has all-white plumage; yellow bill; yellowish to dark green
legs, with duskier feet. Juvenile similar to winter adult except legs
and feet gray-green to blackish.

Similar Species
Great Egret much taller and more slender; longer bill and legs; legs
always black. Snowy Egret has longer, slimmer black bill; yellow feet;
slimmer, larger, and more aquatic. See juvenile Little Blue Heron,
white-phase Reddish Egret, and "Great White Heron."

Range
Breeds in the Southwest to Texas; in the East, in Michigan, from
s. Maine to Florida, and along the Gulf Coast. May wander to Canada
following breeding. Winters in southern parts of range.

Little Blue Heron
Egretta caerulea

Goose-sized
*Freshwater Marshes; Mudflats
and Tidal Shallows; Salt Marshes;
Northern Wooded Swamps;
Southern Wooded Swamps*

Field Marks
24″. A fairly common dark heron of marshes, mudflats, shallows, and
wooded swamps. Feeds by wading slowly through shallow water,
searching quietly for small fishes and mollusks. Flies with neck folded
and feet trailing.
Breeding adult (2) blue-gray with reddish-purple feathering on head
and neck. Nonbreeding adults slate-blue with dark purplish gloss on
neck (sometimes indistinct). Both plumages show bluish bill with dark
tip. Immature (1) white overall; bill pale with darker tip.

Similar Species
Reddish Egret has rufous feathering about neck; bill pink with dark
tip; rare white-phase Reddish Egret distinguished from immature
Little Blue Heron by pink bill and dark legs. Snowy Egret, also
similar to immature Little Blue, has yellow skin about eyes, long
black bill. Tricolored Heron has white belly and rump; dark grayish
neck without purple gloss.

Range
Breeds from New England south to Florida and along Gulf Coast.
May wander farther north. Winters along coast from New Jersey
south to Mexico.

Reddish Egret
Egretta rufescens

Goose-sized
Gulf Coast and Florida: Mudflats;
Inshore Waters; Freshwater
Marshes

Field Marks
30″. A large wading bird of mudflats, inshore waters, and freshwater marshes along Gulf Coast and Florida. Has unique feeding technique, called canopy feeding: spreads wings to create shade, reducing glare and aiding visibility as it forages for small fishes and crustaceans. Often sprints actively through shallows in pursuit of prey. Two color phases; white-phase birds very rare.
Dark-phase adults (2) have reddish head and neck, dark gray body; bill pink with dark tip. White-phase adults (1) white all over, with dark blue legs and pink bill with black tip. Birds in breeding plumage have shaggy plumes on head and neck (bright reddish in dark phase, white in white phase). Immatures grayish, with some reddish tones on head and neck; bill dark.

Similar Species
See Little Blue Heron. Great, Snowy, and Cattle egrets similar to white-phase Reddish Egret, but differ in size, bill color, and leg color.

Range
Breeds locally on Texas coast and in s. Florida. Occasionally wanders north as far as North Carolina.

Great Blue Heron
Ardea herodias

Very Large
All freshwater habitats; Mudflats
and Tidal Shallows; Salt Marshes

Field Marks
46″. The familiar "crane" is really America's largest heron, not a true crane. Tall and heavy; fishes deliberately, watching for prey. Flies with long neck coiled back onto itself and long legs trailing behind; slow wingbeats. Common and widespread on freshwater and saltwater habitats.

Adults (1) mostly medium gray, with narrow plumes over back and breast; crown and face white, with broad black eyebrow extending into long, narrow black plumes from nape; stripe down foreneck expands into black-and-white breast; black belly and patch where wing tucks; thighs rufous; primaries and secondaries dark gray; bill large, yellowish, daggerlike; legs long, grayish. Juvenile lacks plumes, mostly grayish with rufous feather fringes; cap blackish; throat white; upper mandible black. "Great White Heron" (2), a Florida color phase of Great Blue, all-white with yellow bill and pale yellow legs. Call a deep, rough croak.

Similar Species
See Sandhill Crane, juvenile Reddish Egret, Tricolored and Little Blue herons. Egrets smaller than "Great White"; legs dark.

Range
Breeds across s. Canada south to Mexico and Florida. Winters from British Columbia and s. New England southward.

Tricolored Heron
Egretta tricolor

Goose-sized
Salt Marshes; Mudflats and Tidal
Shallows; Freshwater Marshes

Field Marks
26″. Formerly called Louisiana Heron. An unmistakable wading bird of salt marshes, shallows, freshwater marshes, and mudflats along Atlantic and Gulf coasts. Appears slimmer and longer than most other herons. Often wades into fairly deep water to forage.
Adults (1, 2) have dark gray-blue upperparts and neck; belly and rump white; white line runs from breast to chin; throat washed with maroon or chestnut. Breeding adults (1) show maroon feathering along back as well. Immatures have more maroon or reddish feathering along wings and back of neck. Bill in all plumages very long, slender.

Similar Species
Little Blue Heron lacks white on belly. Great Blue Heron much larger, paler overall.

Range
Breeds along Atlantic Coast from New Jersey to Florida, and along the Gulf Coast. Birds in northernmost parts of range move south in winter.

Yellow-crowned Night-Heron
Nycticorax violaceus

Goose-sized
All freshwater habitats; Mudflats
and Tidal Shallows; Salt Marshes

Field Marks
24″. A grayish wading bird with a relatively short neck; found in all
freshwater wetland habitats of the East, as well as in mudflats and
tidal marshes. In flight, feet extend beyond tail. Chiefly nocturnal, but
sometimes forages by day.
Adults (1) mainly gray, with black head, white cheek patches, and
white crown; long white plumes on crown. Immatures (2) brownish
with white spots above and fine streaks below. In all plumages, bill is
stout.
Call a loud *kwawk* or *wawk*.

Similar Species
Black-crowned Night-Heron in flight has only toes extending beyond
tail; call flatter, somewhat lower-pitched than Yellow-crowned's.
Immature Black-crowned has less spotting above; plumage warmer
brown.

Range
Breeds from Minnesota and Ohio south to Gulf Coast, and from New
England to Florida. May wander north to Canada. Winters on
Atlantic and Gulf coasts from North Carolina.

Greater Flamingo
Phoenicopterus ruber

Very Large
Southern Florida: Inshore Waters;
Mudflats and Tidal Shallows

Field Marks
46″. An unmistakable pink wading bird of southern Florida, with
extremely long neck and legs. Flamingos flush seawater through
comblike "teeth" within large, crooked bill, straining out small
creatures; often feed with bill or even entire head submerged in
water. In flight, appears exceedingly long-necked; flocks often fly in
long lines.
Adults (1) rose-pink overall, with long legs and neck; bill heavy,
downcurved, pink at base, with black tip. Immatures much paler pink,
with grayish tones above.

Similar Species
Roseate Spoonbill much smaller, with long, spatulate bill and shorter
neck.

Range
Breeds in West Indies, Galapagos Islands, and Yucatán in Mexico.
Regularly wanders to s. Florida.

Limkin
Aramus guarauna

Goose-sized
Florida: Freshwater Marshes;
Southern Wooded Swamps

Field Marks
26″. A long-necked wading bird, somewhat like a heron in appearance.
A relative of the cranes, it inhabits freshwater marshes and wooded
swamps throughout peninsular Florida. When foraging, skulks along
edges of marshes and ponds, pumping or jerking its tail; uses long bill
to probe muddy bottoms for snails and other mollusks.
Adults (1) brownish overall, with white spots and streaks, especially
on head and neck; bill long, somewhat downcurved, and sturdy; legs
long. Immatures resemble adults, but may be slightly paler.
Call a loud, wailing *kurr-ee-ow* or *krr-ow*; given mainly at night.

Similar Species
Immature Yellow-crowned and Black-crowned night-herons have
shorter, stouter bills, shorter necks and legs. Immature ibises have
much longer, strongly curved bills, lack white spots and streaks.

Range
Resident in se. Georgia and throughout most of Florida. Also in
tropics.

Glossy Ibis
Plegadis falcinellus

Goose-sized
Salt Marshes; Mudflats and Tidal
Shallows; Freshwater Marshes

Field Marks
23″. A dark wading bird of marshes, mudflats, and shallows; locally
common, but range has been expanding. Feeds by foraging in wetland
habitats, using its long downcurved bill to probe for crayfish and other
bottom-dwellers.
Adults in breeding plumage (2) mainly dark chestnut, but washed
with glossy purple; appear all-dark from a distance; face has pale
border. Winter adults brownish, washed with metallic green; head
and neck have white streaks. Immatures resemble winter adults, but
lack any chestnut feathering. In all plumages, brown eye is a good
field mark at close range.

Similar Species
Adult White-faced Ibis, *Plegadis chihi* (1), of coastal Texas and
Louisiana, has white feathering around bare face patch; immatures of
two species not safely distinguishable in field.

Range
Breeds from Maine south along Atlantic Coast to Florida, and along
Gulf Coast. Winters in southern parts of range.

White Ibis
Eudocimus albus

Goose-sized
All freshwater habitats; Mudflats
and Tidal Shallows; Salt Marshes;
Open Country

Field Marks
25″. A white wading bird with a very long, downcurved bill; found in
many different coastal wetland habitats, but also in fields and other
open areas. Flies with neck and legs extended, in long lines or in
V-shaped formations.
Adults (2) white with long, downcurved red bill and patch of red skin
around eye; black wing tips visible in flight (inset). Immatures (1)
brownish above, with white belly; bill somewhat duller and may
appear brownish.

Similar Species
Immature Glossy and White-faced ibises lack white on belly. Limpkin
has white spots and streaks. Immature night-herons have short, stout
bills.

Range
Resident along Atlantic Coast from North Carolina south to Florida,
and along Gulf Coast to Texas. May wander farther north in interior
after breeding season. Also in tropics.

Wood Stork
Mycteria americana

Very Large
Salt Marshes; Mudflats and Tidal
Shallows; Freshwater Marshes;
Southern Wooded Swamps

Field Marks
40″. A very tall, heavy wading bird of southern wetlands. Catches
small fish by feeling with its bill in shallow, often murky, water. Nests
colonially on trees in swamps; always gregarious. Flies with slow,
strong wingbeats, with neck and legs extended. Soars high and far,
using long and broad wings with rounded, "fingered" tips.
Adults (1–3) white with glossy black primaries, secondaries, and tail;
head and thick neck black, naked; neck has dull yellowish scales, and
head has dull yellowish cap; bill long, massive at base, tapering and
downcurved toward tip, dusky horn colored; long legs dark gray; feet
pinkish. Immature similar but has dingy gray feathering on neck and
nape; bill pale yellow.

Similar Species
White Ibis has slender, strongly downcurved bill; much smaller; adult
has black only on small part of wing tip; fast flapping flight. Great
Egret and "Great White Heron" smaller, all-white; hold neck in
S-curve, especially in flight. American White Pelican stockier, with
short neck, massive yellow or orange bill; short legs; white tail.

Range
Nests in Florida and occasionally s. Georgia. Summer visitor to Salton
Sea in s. California, and locally in the Southwest and southern
Mississippi Valley. Occasionally seen north to Great Lakes region and
New England.

146

Roseate Spoonbill
Ajaia ajaja

Goose-sized
Gulf Coast and southern Florida:
Salt Marshes; Freshwater
Marshes

Field Marks
32″. A distinctive pink wading bird with a long, flat, spoon-tipped bill.
Found in saltwater and freshwater marshes along Gulf Coast and in
southern Florida. Consumes small fishes, crabs, and shrimps; feeds by
moving bill slowly from side to side through water, straining out its
prey. Flies with head and neck extended.
Adults (2,3) appear all-pink at a distance; wings and lower back pink,
with bright red wing coverts; upper back, neck, and upper breast
white; tail orange; head naked, and pale greenish-white; legs bright
pink. Immatures (1) pale pink to whitish overall, with feathered head.
Bill distinctive in all plumages.

Similar Species
Greater Flamingo much taller, with short, downcurved bill and
extremely long neck and legs.

Range
Breeds and winters along Gulf Coast in Texas and Louisiana, and in
s. Florida. Also in tropics.

147

Sandhill Crane
Grus canadensis

Very Large
Open Country; Freshwater
Marshes; Salt Marshes

Field Marks
41″. A very large, long-legged, gray wading bird of open country and
freshwater marshes. Sandhills travel in large, noisy flocks; loud,
hollow call, less melodious than call of Whooping Crane, carries for
more than a mile.
Adults (1) gray overall, with bare red patch on crown; neck long,
straight; legs black; females usually somewhat smaller than males.
Populations nesting in wetlands with iron-rich mud may appear
reddish-brown in spring from oxidization of mud on feathers.
Immatures reddish-brown; lack bare red spot on crown.
Gives a long, hollow, rattling *garoooooooooo*; flocks often call in flight;
male and female also give a "duet."

Similar Species
See immature Whooping Crane.

Range
Breeds from n. Alaska and across Canada to Hudson Bay. Migrates
to Rocky Mountains and Great Lakes region. Winters in central
California and through the Southwest to Texas. Also resident in
s. Florida.

Whooping Crane
Grus americana

Very Large
Salt Marshes; Freshwater
Marshes

Field Marks
52″. An extremely rare, very large white wading bird of freshwater marshes and open-country wetlands. Unmistakable in flight, with head and neck held straight out, feet trailing behind, and wingspan of more than seven feet; wings flap slowly down, then more rapidly upward. Parties of three or four feed together, usually taking small fishes, crustaceans, and mollusks.
Adults (1) white with black wing tips; neck long, straight; bare patch of red skin on crown and along jaw; feet and legs dark. Females usually somewhat smaller than males. Immatures washed with cinnamon or rust; white on belly and wings.
Gives a musical *ker-loo, ker-loo*; may carry for more than a mile. Male and female also give a "duet."

Similar Species
American White Pelican flies with neck tucked in; appears short-legged. Snow Goose much smaller. In good light, Sandhill Crane's gray plumage easily distinguished; Sandhill also smaller.

Range
Nests in west-central Canada and migrates to Aransas National Wildlife Refuge on coast of Texas. New population nests in Rocky Mountains of Idaho and migrates to New Mexico.

Long-billed Curlew
Numenius americanus

Crow-sized
Mudflats; Beaches; Salt Marshes;
Open Country and Grasslands

Field Marks

23″. Tall and long-necked; our longest shorebird, with an extremely long, downcurved bill. Probes mudflats and sand flats for prey, often twisting neck and bill to follow worm burrows. Solitary or in flocks on tidelands; also in flocks in open country inland. Breeds on inland grasslands.

Adults (1) warm buff-brown, paler below; heavily dotted with buff on brown above and on wings; finely streaked on neck and breast; crown streaked brown and buff, but without bold stripes; cinnamon underwings visible in flight (inset); bill extremely long, fine, and strongly downcurved; especially long in female; legs long, grayish. Juvenile similar but with much shorter bill well into first fall; buffier with less streaking on breast and throat.

Common call is a clear, rising, whistled *cur-lee*. Short alarm trill.

Similar Species

Whimbrel grayish-brown, including underwing; proportionally stockier, with generally shorter bill; smaller; head has bold stripes. Other large shorebirds lack strongly downcurved bill. Ibises heavier, with thicker bill and unpatterned back.

Range

Breeds from s. Canada south to ne. California and through the Southwest to Texas. Winters from s. California to n. Texas and Louisiana, and from South Carolina south to Florida.

Whimbrel
Numenius phaeopus

Crow-sized
Beaches; Mudflats; Salt Marshes;
Freshwater Marshes;
Open Country and Grasslands

Field Marks
17½". A large, tall brown shorebird with a long, strongly downcurved bill, used in probing for prey. Rather long-necked. Usually in small flocks on mudflats, rocky shores, and beaches, or in larger flocks on migration, when it visits lakes and plowed fields. Breeds on arctic tundra.
Adults (1) grayish-brown above and on wings, dappled with paler spots; whitish below, finely streaked with brown on neck and breast; barred flanks; crown strongly striped, with broad lateral crown stripes and eyelines dark brown against buff-white eyebrows and median crown stripe; long, strongly downcurved bill; long gray legs. Juvenile similar, but with shorter bill into first fall; slightly crisper, dull buffy spots above and finer streaks on breast.
Call a fast bubbling trill of uniform, short whistled notes; carries far.

Similar Species
Long-billed Curlew warm buff-brown; usually with longer, thinner bill; larger; cinnamon underwing; less distinct head stripes. Other large shorebirds lack strongly downcurved bill. Ibises larger and heavier, with thicker bill and unpatterned back.

Range
Breeds in Alaska and n. Canada. Winters along Pacific and Atlantic coasts from California and North Carolina south. Migrant along Gulf Coast. Sometimes wanders inland.

151

Marbled Godwit
Limosa fedoa

Crow-sized
Beaches; Mudflats; Salt Marshes;
Freshwater Marshes; Ponds; Open
Country

Field Marks
18″. A large, tall, long-necked, buff shorebird, with a long, slightly upcurved bill. Wades along water's edge, probing vertically into mud, sand, or beach wrack for small prey. Breeds on prairie potholes. Migrates through various inland wetlands.
Breeding adults buff and brown; heavily dotted with buff on brown above and on wings; dull cinnamon underwings; underparts warm buff; finely streaked brown on neck and head; rest of underparts except lower belly thickly covered with fine brown bars; very long, thin, slightly upcurved bill, pink with dark tip; long grayish legs.
Winter adults (1) and juveniles similar, but underparts lack bars and may fade to whitish.
Typical calls include single or repeated, rising *ker-whit!* and loud, chanting *radica, radica, radica.*

Similar Species
Hudsonian Godwit has strong white wing stripe and rump, black wing linings and tail band; smaller. Long-billed Curlew and Whimbrel have strongly downcurved bill. Willet and Whimbrel grayer; Willet has shorter, thicker bill, black-and-white wing pattern, and white rump.

Range
Breeds across central Canada south to Montana, South Dakota, and Minnesota. Winters along coasts from California to Mexico and from Virginia (rare) south; rarely on Gulf Coast.

152

Hudsonian Godwit
Limosa haemastica

Crow-sized
Mudflats; Beaches; Salt Marshes;
Freshwater Marshes; Ponds

1

Field Marks

15½". A very long-billed shorebird that nests in the Far North; commonly observed in our range only during spring migration from wintering grounds in South America. Flocks feed along beaches, mudflats, and in marshes; these birds use long bills to probe shoreline for small marine creatures.

Adults in breeding plumage (1) mottled brownish-black above, chestnut below with some black-and-white barring. Winter adults paler gray-brown overall, with whitish lower breast and belly. In both plumages, bill long, slender, slightly upturned; pink at base, with black tip. In flight, both plumages show black wing linings, white wing stripe, and black band on white tail. Juveniles dark brown above, paler buff-brown below; rarely encountered.

Similar Species

Marbled Godwit larger; lacks white wing stripe and black band on tail; shows cinnamon-red wing linings in flight.

Range

Breeds in Alaska, British Columbia, and near Hudson Bay. Winters in South America. Migrates north across the Great Plains and south along Atlantic Coast, mainly offshore.

153

Black-necked Stilt
Himantopus mexicanus

Crow-sized
Mudflats and Tidal Shallows; Salt
Marshes; Lakes and Ponds

Field Marks
14″. The extraordinarily long, thin legs of this elegant shorebird fit its
name. Wades in deeper water than most shorebirds, where it picks
small invertebrates off surface. Frequents salt ponds; less often seen
along tidal shores. Locally abundant.
Adult male (1) glossy black on back, wings, hindneck, crown, and ear;
has white underparts, foreneck, and face, white spot over eye, white
rump, and wedge up lower back; pale gray tail; legs stiltlike, pink-red;
bill nearly straight, needlelike, black. Adult females similar, but with
brown back and brownish-black plumage from hindneck through
crown. Juveniles like females, but upperparts fringed golden buff,
fading and wearing thin as bird grows.
A loud, yapping *pip-pip-pip-pip-pip*, becoming incessant when
excited.

Similar Species
American Avocet has broad white stripes on back and wings; head
and neck pale or rufous; upswept bill. Other tall shorebirds lack
extraordinarily long pink-red legs, black-and-white pattern.

Range
Breeds locally along Pacific Coast from Oregon south, Atlantic Coast
from Delaware south. Scattered inland locations in the West. Winters
from s. Oregon south and in s. Florida. Also on Gulf Coast.

154

American Avocet
Recurvirostra americana

Crow-sized
Mudflats and Tidal Shallows; Salt
Marshes; Lakes and Ponds

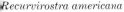

Field Marks
18″. Conspicuous, gregarious, and easily identified. A tall shorebird
with distinctive foraging technique: Sweeps its distinctively upcurved
bill quickly from side to side through water or the surface of soft mud
to trap tiny food items. Walks, wades, or swims. Numerous at salt- or
freshwater lakes and ponds, marshes, and muddy tidal shores.
Breeding adult male (2) has broad blackish and white stripes on back
and inner half of upper wing; primaries black; belly and tail region
white; head, neck, and breast rufous-buff; long, thin gray legs; black
bill very long, very fine, distinctly upswept. Female similar, but bill
more strikingly upcurved. Winter adults (1) have rufous-buff replaced
by pale gray, darkest on hindneck. Juvenile resembles adult, but has
rufous-buff hindneck and dingy grayish crown; browner back.
Broadcasts loud, repeated *kleek* calls.

Similar Species
Black-necked Stilt has solid blackish back and wings; straight bill;
black-and-white neck and head; pink-red legs. Godwits and Willet
have thicker, straighter bills; lack white back stripes.

Range
Breeds locally on California coast, inland to w. Minnesota and Texas.
Winters along Pacific Coast from California south, and inland at
Salton Sea; on Atlantic Coast, from North Carolina south; and on Gulf
Coast. May wander farther north.

American Oystercatcher
Haematopus palliatus

Crow-sized
Beaches; Mudflats

Field Marks
18½". A distinctive black-and-white shorebird with a long, straight red bill. Found along beaches and mudflats, in family groups during nesting season or in small flocks at other times of year. Feeds on clams, mussels, and other bivalves, in addition to oysters, prying them open with chisel-like bill.

Adults (1) dark brown above, with white belly; head and neck black; eyes yellow, with red outline; bill bright red or reddish-orange; in flight, white wing patches and upper-tail coverts contrast boldly with dark back and wing tips. Immatures duller; bill orange with dark tip; eyes dark.

Call a ringing *wheep* or *cleep*, often repeated; in flight, gives a *kreek* or *crik*.

Range
Breeds from Massachusetts south along Atlantic Coast to Florida, and on Gulf Coast from Texas to Louisiana. Winters north to Delaware. May wander farther north.

Ruddy Turnstone
Arenaria interpres

Robin-sized
Beaches; Rocky Shores; Mudflats

Field Marks
9½″. A gregarious shorebird; abundant on pebble beaches and beach
wrack, but also uses rocky shores, sandy beaches, mudflats, and,
occasionally, uplands. Flips over stones and pulls apart rotting kelp.
Plump; short-legged.
Breeding adults (1) have complex black-and-white harlequin face and
neck pattern; black breast with white at upper sides; white of flanks
and belly extends up as point onto middle of breast; back, scapulars,
and wing coverts boldly patterned in rufous and black, with white
down middle of back and on inner coverts; rump white; tail black
except narrow white tip; black bill short and slightly upcurved, tapers
to fine point; short legs orange-red. Winter-plumage birds similar in
pattern but much duller; brown above and on head and neck, buff
where white in summer (except white throat); black breast mottled.
Juveniles like winter adults but even duller; browner, with broad buff
fringes above.
Call a low, mechanical rattle, most often given in flight.

Range
Breeds from Alaska across n. Canada. Winters on Pacific Coast from
Oregon south, along Atlantic Coast from Connecticut south; also along
Gulf Coast. Rare inland.

American Woodcock
Scolopax minor

*Robin-sized
Northern Wooded Swamps;
Southern Wooded Swamps;
Eastern Deciduous Forest;
Thickets*

Field Marks
11″. A shy, stocky, long-billed shorebird of thickets, wooded swamps, and wet forested areas. Easiest to observe in spring, when males perform spiraling courtship flight, calling and circling high as wings make a loud whistling noise.
Adults (1) mottled with buff and black above; wings dark brown, finely streaked with buff; crown dark, with four rust-colored bars; underparts bright rust; bill very long, slender; eyes dark, set very far back in head; in flight, wings appear rounded. Immatures resemble adults.
Gives a nasal, buzzy *bzeep* or *peeent*; male gives chirping whistles in courtship flight.

Similar Species
Common Snipe has more pointed wings; plumage much more striped with dark brown and black.

Range
Breeds across se. Canada south to e. Texas and Florida. Winters from the Southeast as far north as New England (rare).

Common Snipe
Gallinago gallinago

Robin-sized
Freshwater Marshes; Ponds

Field Marks
10½″. A common and widespread, but solitary and secretive
inhabitant of wet grassy swales and marshes. Occasionally seen
feeding near dowitchers. Very compact build and short legs, but
extremely long, straight bill. Flushes with startling abruptness and
scratchy call, erratic flight.
Adults (1) have crown, back, and scapulars black-brown with bold
buff-white stripes and rufous-buff dots and lines; brown eyestripe and
whisker stripe parallel crown stripes; underparts buff to white; neck
and breast streaked with brown; sides, flanks, and undertail coverts
barred with brown; brown wings have pale buff covert fringes; tail
rounded, with rufous band near tip; bill about twice length of head.
Flushing call scratchy *skrrrt*. During shallow dives of circling display
flight, tail produces an eerie, pulsing *huhuhuhuhuhuhu!* Ground call
during breeding season a slow, flickerlike *wheek-a-wheek-a-wheek-a*.

Similar Species
Dowitchers not so striped above and on head; have longer legs and
white wedge up back; barred tail; gregarious and usually in open.
American Woodcock plumper, with short, rounded wings; rufous
below, barred above.

Range
Breeds from n. Alaska across Canada, south to California and New
Jersey. Winters from British Columbia and Virginia south.

Long-billed Dowitcher
Limnodromus scolopaceus

Robin-sized
Mudflats; Freshwater Marshes;
Salt Marshes; Ponds

Field Marks
11½". A gregarious, abundant, long-billed shorebird of ponds and
marshes. More partial to freshwater habitats and marshes than the
nearly identical Short-billed Dowitcher, but both feed on tidal
mudflats. Unlike Short-billed, also winters inland. Medium-length
yellow-green to gray legs.
Breeding adults (1) have rufous underparts, densely spotted with
black on throat and breast, and barred with black on sides, flanks, and
undertail coverts; upperparts blackish, with rufous edges and bars;
tail and rump densely barred; white wedge up middle of back. Winter
adults (2) brownish-gray above and from head to breast; white belly;
eyestripe dusky and eyebrow whitish; usually lacks specks on breast.
Juvenile black-brown above, with rufous restricted to narrow edges;
underparts grayish-buff with many faint speckles.
Typical call a sharp, high *keek!*, but any number of notes given.

Similar Species
Short-billed Dowitcher's call a *tu-tu-tu*; bill averages shorter; juvenile
brighter buff, has rufous marks within tertials and scapulars; breeding
has spotted sides, unspotted throat, and/or white belly. See Red Knot
and Common Snipe.

Range
Breeds in Alaska and nw. Canada. Winters on Pacific, Atlantic, and
Gulf coasts from Washington and Virginia south.

Short-billed Dowitcher
Limnodromus griseus

Robin-sized
Mudflats; Salt Marshes;
Freshwater Marshes; Ponds

Field Marks

11″. Sometimes called the "sewing machine" of mudflats, probing for food with a stitching motion of its long bill. Abundant and gregarious. Principally coastal, but also migrates through interior. Medium-length yellow-green to gray legs.

Breeding adults (1) have mostly rufous underparts, with variable white from belly to undertail, and variable black spots from neck through sides and flanks; flanks often barred; upperparts blackish with rufous edges and bars; tail and rump densely barred; white wedge up middle of back. Winter adults brownish-gray above and from head to breast; white belly; dusky eyestripe and whitish eyebrow; normally faint dark specks on breast. Juveniles (2) black-brown above, with broad rufous-buff edges; rufous-buff spots and lines within dark tertials and scapulars; most of underparts buff with many faint speckles.

Typical call a low, mellow *tu-tu-tu*; but any number of notes given.

Similar Species

Long-billed Dowitcher's call a *keek!*; bill averages longer. Juvenile darker and grayer, with only narrow rufous edges above; breeding adult has barred sides, densely spotted throat, and rufous belly (unless molting). See Red Knot and Common Snipe.

Range

Breeds locally from s. Alaska across n. Canada to Labrador. Winters on coasts from n. California and North Carolina south.

Greater Yellowlegs
Tringa melanoleuca

Pigeon-sized
Salt Marshes; Mudflats; All
freshwater habitats

Field Marks
14″. A tall, slender, fairly active wading shorebird. Usually solitary or in small flocks. Often feeds by sweeping bill back and forth in water. Visits all freshwater habitats, marshes, and mudflats.
Breeding adults (1) blackish-brown above, densely dotted with white; head, neck, and breast finely streaked dark on white; flanks barred; belly and rump mostly white; long, thin legs bright yellow to orange; bill long, thin, slightly upcurved, black with grayish base. Winter adults (2) brownish-gray above, with some fine white specks; white below, with fine gray-brown streaks on breast and neck; flanks lightly barred. Juveniles like winter birds but darker brown above, with more contrasting whitish specks.
Typical call a loud, piercing, slightly descending *teu-teu-teu!*

Similar Species
Lesser Yellowlegs smaller, with shorter, thinner, straight black bill; less strongly marked in most plumages; has shorter, weaker call. Willet has gray legs; black-and-white wings. Ruff, *Philomachus pugnax*, a rare visitor from Eurasia, is stockier, with narrow white wing stripe, shorter bill, and duller legs. See Solitary Sandpiper.

Range
Breeds from s. Alaska across central Canada to Labrador. Winters along coasts from Washington and New York south; and inland across southernmost states. Common migrant in much of United States.

162

Lesser Yellowlegs
Tringa flavipes

Robin-sized
Freshwater Marshes; Salt
Marshes; Mudflats; Ponds

Field Marks
10½″. A long-legged, rather small, very active wading shorebird.
Often in large flocks in East, but scarcer in West. Prefers ponds and
marshes; sometimes seen on mudflats.
Breeding adults (1) blackish-brown above, densely dotted with white;
head, neck, and breast finely streaked dark on white; flanks lightly
barred; belly and rump white; legs long, very thin, bright yellow; bill
thin, relatively short, straight; nearly all-black. Winter adults
brownish-gray above, with tiny whitish specks limited to wings; white
below, with very fine gray streaks on breast and neck. Juveniles (2)
dark brown above, with buff-white specks; white below, with blurry
gray streaks on breast and neck.
Typical call a *tew-tew* or *tew*, higher and weaker than call of Greater
Yellowlegs. Sometimes gives three or more notes.

Similar Species
Greater Yellowlegs larger, with longer, thicker, slightly upcurved
gray-and-black bill; more strongly marked in most plumages; gives
longer, louder call. Solitary Sandpiper has dark rump; shorter
greenish legs.

Range
Breeds from Alaska across central Canada to southern Hudson Bay.
Winters on Pacific Coast from s. California, and along Atlantic and
Gulf coasts from Virginia.

Willet
Catoptrophorus semipalmatus

Crow-sized
Salt Marshes; Mudflats; Beaches;
Freshwater Marshes

Field Marks
15″. A tall, robust, grayish shorebird. Uses straight, fairly thick bill to probe for a variety of invertebrates; patrols mudflats, beaches, rocky shores, and marshes; wades in shallow water. Breeds in prairie potholes and eastern salt marshes. Often in large flocks.
Breeding adults (2) gray above, heavily spotted with dark brown; whitish below, neck and head finely streaked brown, breast and flanks finely barred brown; wings largely black, with striking broad white stripe through bases of flight feathers; rump white; tail gray; bill medium length, straight, fairly thick, grayish-black; long legs gray. Winter adults (1) plain brownish-gray above; pale gray with tiny streaks on neck, breast, and flanks. Juveniles similar to winter adults, but finely peppered with buff and brown flecks throughout.
Gives repeated whistled calls: *pill-will-willet!*, *weep-weep-weep!*, and others.

Similar Species
Greater Yellowlegs has yellow legs; plain wings; thinner bill. Godwits have longer, thinner, pink-based bill. Marbled Godwit buff with plain wings; larger.

Range
Breeds from central Canada south to ne. California, Nevada, and Nebraska, and in East, along Atlantic and Gulf coasts from Nova Scotia south. Winters along coasts from Oregon and North Carolina.

Upland Sandpiper
Bartramia longicauda

Pigeon-sized
Open Country and Grasslands

Field Marks
12″. A nearly solitary inhabitant of tall grasslands, breeding on
prairies and fallow fields; also uses other uplands. Unique shape
makes it easy to recognize: has short, thin bill, small head atop long
neck, and comparatively long tail; legs rather long. Lands on posts
and poles, often holding wings up briefly.
Adults (1) dark brown above and on upper wings, with feathers
fringed golden-buff and barred with black; streaked crown with
narrow buff median stripe; throat whitish; face, side of neck, and
breast buff, with fine dark streaks becoming larger chevrons on lower
breast and flanks; belly whitish; small head and buff eye-ring
accentuate large dark eye; small, pale bill; rather long, pale yellow
legs. Juveniles similar, but with more scaly buff fringes above and less
barred flanks.
Unique song begins with short rattle, continues with two long,
haunting whistles, first sliding up pitch, second sliding back down.
Flight call a *pip-pip-pip-pup.*

Similar Species
Buff-breasted Sandpiper smaller and more compact, without strong
markings below. Curlews larger, with downcurved bill.

Range
Breeds from Alaska east to New Brunswick and south to Oregon, n.
Oklahoma, and Virginia. Winters in South America.

Stilt Sandpiper
Calidris himantopus

Robin-sized
Ponds; Mudflats; Freshwater
Marshes; Salt Marshes

Field Marks
8½". A very long-legged, slim shorebird with a long, slightly
downcurved bill. In our range, seen chiefly in migration, when it
frequents ponds, mudflats, and freshwater and salt marshes; flocks
sometimes include hundreds of birds. Feeds by probing rapidly and
repeatedly into mud for small crustaceans and mollusks.
Breeding adults (1) have streaked crown, mottled upperparts, and
white underparts with heavy pattern of barring and streaking; some
chestnut stripes on sides of head; in flight, white rump and tail visible.
Winter adults paler gray-brown overall, with faint spotting and
streaking. Juveniles (2) resemble winter adults, but have conspicuous
buff edgings to wings; breast and neck pale buff with some faint
streaking. All plumages have a white eyebrow.
Gives a *werp* or *querp* call in flight.

Similar Species
Dowitchers heavier-looking, with longer bills; have unspotted belly in
breeding plumage; show white wedge on back in flight. Lesser
Yellowlegs has bright yellow legs, straighter bill.

Range
Breeds in n. Alaska and arctic Canada. Winters mainly in South
America, but a few winter in s. Florida, along Gulf Coast of Texas,
and at Salton Sea, California. Migrates over the Great Plains and
occasionally along Atlantic Coast.

Dunlin
Calidris alpina

Robin-sized
Mudflats; Beaches

Field Marks

8½". Arrives later in fall than other shorebirds; winters farther north than most. Abundant on mudflats, common on beaches, also uses rocky shores. Distinctive bill a bit longer than head, distinctly drooped at tip.

Breeding adults (1) have back, scapulars, and crown chestnut, streaked with black; belly black; flanks and undertail coverts white; breast, neck, and face white with fine dark streaks; upper wing, tail, and center of rump gray-brown; wing stripe white; bill and legs black. Winter adults (2) plain brownish-gray above and on head, neck, and breast; variable dusky streaking within gray breast; remainder of underparts white. Juveniles have black back with chestnut and white fringes; rest of upperparts brown with buff fringes; lines of dark spots on white flanks and sides of belly; breast washed buff and finely streaked; face buffy gray-brown.

Gives a grating *greep* in flight. Soft twitters from flock on ground.

Similar Species

Nonbreeding Western and Semipalmated sandpipers smaller with shorter bills; breast white or pale buff; paler gray above in winter. Winter Sanderling whiter, with straight bill. See Purple Sandpiper.

Range

Breeds in n. Canada south along Hudson Bay. Winters from s. Alaska south on Pacific Coast; Atlantic and Gulf coasts from Massachusetts.

Black-bellied Plover
Pluvialis squatarola

Robin-sized
Mudflats; Beaches; Salt Marshes;
Freshwater Marshes; Open
Country

Field Marks

11½″. The largest plover; most numerous on mudflats, but also uses
beaches, marshes, inland shores, and fields. Most frequently seen
along coasts in winter. Forages in typical look-run-look-peck plover
style. Heavy plover shape, with short, thick black bill. Always has
black axillars, white rump, barred tail, and grayish wings with
white stripe.
Breeding males (2) pure black on face and underparts, with white
undertail coverts, sides, and border of face; upperparts black-brown
with many large white spots. Breeding females similar but duller;
brownish-black below mottled with white; browner above with less
white. Winter adults (1) brownish-gray above with whitish fringes;
white below with gray-brown smudges on breast. Juveniles black-
brown above with many small spots that start out yellowish and soon
fade to white; otherwise like winter birds, but finely streaked from
throat to legs.
Gives a clear, whistled *whee-ur-eee*, middle note lower; carries far.

Similar Species

Lesser Golden-Plover smaller, with smaller bill and dark rump; lacks
white wing stripe and black axillars; usually more golden above.

Range

Breeds in n. Canada. Winters along coasts from British Columbia and
Massachusetts southward.

168

Lesser Golden-Plover
Pluvialis dominica

Robin-sized
Open Country; Freshwater
Marshes; Salt Marshes; Mudflats

Field Marks
10½". A large plover, nearly always speckled golden-yellow above, with plain dark wings and rump. In migration, most often seen in plowed fields and open country. Often holds wings over head briefly on alighting.
Breeding males (2) pure black on face and underparts; broad white borders of face and side of neck converge slightly on sides of breast; black above, heavily speckled with gold. Breeding females similar but browner below, with some white mixed in. Winter adults similar but browner, with gold specks often reduced or even absent; breast smudged gray-brown. Juveniles (1) black-brown above, heavily speckled with gold (often pale); face through breast and flanks heavily but finely speckled pale gray-brown on whitish background.
Gives a plaintive, rising *twee* or *tu-wee*, with little or no dip in middle.

Similar Species
Black-bellied Plover larger, with larger bill; white rump and wing stripe; black axillars; only briefly golden-speckled above.

Range
Breeds in Alaska and n. Canada. Migrates in spring across Great Plains; in fall, inland and along Atlantic Coast, less abundantly along Pacific Coast. Winters in South America.

Killdeer
Charadrius vociferus

Robin-sized
Open Country; Mudflats; Lakes,
Ponds, and Rivers

Field Marks
10½". Technically a "shorebird," but more a bird of open uplands;
common on almost any grassy or open ground, including fields and
lawns. Also visits both fresh- and saltwater shores. Prefers to nest on
gravel near water. Vociferous. Forages night or day.
Adults (1) have white underparts and underwings with two
contrasting broad black bands across breast; white collar completely
encircles neck over upper black band, which reaches nape; crown,
back, scapulars, tertials, and wing coverts plain brown; rump bright
rufous; tail long and tapered, brown and rufous, with white edges and
black subterminal band; upper wing with bold white stripe; blackish
forecrown and face mask separated by white forehead and eyebrow;
bill small and black; eyes large and dark. Juveniles similar but duller,
with down strands often lingering on tail tip.
Piercing alarm whistles are familiar sound, day or night. Gives its
name: *kill-dee!* Also *dee-dee-dee!* or trilled *tree-dididididee!*

Similar Species
Other "ringed" plovers all smaller; have only one complete or partial
black ring.

Range
Breeds from Alaska to Newfoundland and south through most of n.
United States. Winters throughout s. United States and north along
coasts to British Columbia and New Jersey.

Wilson's Plover
Charadrius wilsonia

Robin-sized
Beaches; Mudflats

Field Marks
7¾". A small, comparatively tame plover of beaches, sand dunes, and mudflats. Nests in a hollow depression, or scrape, like most members of its family; will try to distract intruders by pretending to create a new nest in some other location. Feeds on small crustaceans and mollusks; usually seen singly or in pairs. Searches for food by running and stopping.

Breeding males (1) dull brown above, with black breast band, white throat, forehead, and eyebrow; bill heavy, black, comparatively long; breast and belly white. Winter males and adult females (2) similar but paler; breast band brown, not black. Immatures resemble females but may have incomplete breast band.

Similar Species
Semipalmated Plover somewhat darker above, with narrower breast band, short bill with pink base, black tip. Piping Plover much paler, with black line across forehead and much shorter bill with pink base. Killdeer larger, darker above, with two black breast bands and short, dark bill.

Range
Breeds along coasts from s. California and Maryland southward. Winters along Florida coasts, occasionally farther north, and in s. California.

Semitalmated Plover
Charadrius semipalmatus

Sparrow-sized
Mudflats; Beaches

Field Marks
7¼". A small "ringed" plover found mostly on mudflats and sandbars,
but also on beaches. Like other plovers, defends individual space for
its look-run-look-peck foraging, but gregarious at other times. Very
compactly built, with tiny bill.
Breeding adults (2) have underparts and underwings entirely white
with single broad black band across breast; white from throat
completely encircles neck; upperparts and wing coverts almost
entirely plain brown; tail rounded, with white edges and dark brown
subterminal band; upper wing has bold white stripe; white forehead
surrounded by black forecrown and blackish face mask; bill tiny,
orange with black tip; eyes large and dark; legs orange. Winter adults
similar, but black is replaced by brown, and bill duller. Juveniles (1)
like winter adults, but minutely fringed with buff above; bill nearly
all-black; legs dull yellow.
Flight call a distinctive, enunciated *chu-wee*.

Similar Species
Wilson's Plover has much larger black bill; gray-pink legs. Piping
Plover pale gray above; rump white. See Killdeer and Snowy Plover.

Range
Breeds from Alaska to Nova Scotia. Winters along Pacific Coast from
California south and along Atlantic and Gulf coasts from South
Carolina. Migrates along coasts and inland.

172

Piping Plover
Charadrius melodus

Sparrow-sized
Beaches; Rivers; Mudflats

Field Marks
7¼". An uncommon inhabitant of sandy beaches, mudflats, and
sandbars along large rivers. Most often seen along water's edge,
running, stopping deliberately, seizing food, and running on. Calls
often, but difficult to observe when crouched against backdrop
of sand.
Breeding adults (2) sandy gray above, with black collar and black
stripe between eyes; forehead, throat, and underparts white; white
rump visible in flight; bill small, stubby, orange with black tip; feet
and legs orange. Winter adults (1) lack breast band and stripe on
head; bill black; feet and legs dull orange. Juveniles similar to
winter adults.
Call a piping *peep* or *peep-lo*.

Similar Species
See Wilson's Plover. Semipalmated Plover has much darker
upperparts, dark rump. Snowy Plover smaller, darker; has black
patches (not band) on breast; longer, dark legs; lacks orange on bill in
breeding birds.

Range
Breeds locally from Newfoundland south to North Carolina and from
central Canada to Great Lakes region. Winters on Atlantic Coast from
North Carolina south; also on Gulf Coast.

Snowy Plover
Charadrius alexandrinus

Sparrow-sized
Beaches

Field Marks
6¼". A tiny, pale gray-brown plover, cryptic on its habitat of sandy beaches and salt flats. Has running-and-stopping foraging style, typical of plovers. Compact build, but slender for a plover, with thin, short black bill.

Breeding males (1) have all-white underparts and underwings, except broad black bar on each side of breast just in front of wing; upperparts and wing coverts almost entirely plain, pale brownish-gray; tail and rump with white sides; upper wing with fairly bold white stripe, visible in flight; face white, with black forecrown and ear patch; legs dark gray. Breeding females similar, but black on head and sides of breast replaced by black-brown to brownish-gray. Winter-plumage adults similar to dull breeding female. Juveniles (2) like winter adult, but with fine buff fringes above; legs often pale grayish.

Calls include low *krut* and soft *ku-wheet*.

Similar Species
Piping Plover heavier; orange to yellow legs; shorter, thicker bill, often orange-based; paler. Semipalmated and Wilson's plovers have complete breast band; darker brown above; heavier; bill thicker.

Range
Resident locally along Pacific Coast, in Southwest, on Great Plains, and along Gulf Coast.

Wilson's Phalarope
Phalaropus tricolor

Robin-sized
Freshwater Marshes; Salt
Marshes; Mudflats; Lakes and
Ponds

Field Marks
9¼". Better suited to wading than swimming, this tall phalarope is
not pelagic. Breeds in marshy potholes, but visits other freshwater
habitats and estuaries. Needlelike bill slightly longer than head.
Breeding females (2) gray above, with chestnut stripes on back; crown
to upper back whitish gray; black stripe through eye and down side of
neck, ending at chestnut patch; broadly rufous-buff across breast and
neck; throat and rest of underparts white; gray-brown wings without
stripe; rump white; tail gray; legs black. Breeding males smaller;
gray-brown above and white below, with only a trace of female's
colors. Winter adults plain brownish-gray above with white rump;
pure white below; dark eyestripe and medium gray crown contrast
weakly with white face. Juveniles (1) like winter adult, but darker
brown above with buff fringes; buff wash on side of breast; yellowish
legs.
Call a low, nasal *wurk, wurk, wurk.* Also toots when breeding.

Similar Species
Other phalaropes have white wing stripe; dark-centered rump;
compact. See Stilt Sandpiper and Lesser Yellowlegs.

Range
Breeds from sw. Canada south to central California, Great Plains, and
Great Lakes region. Winters in South America, occasionally in
California and Texas. Migrates mainly across Great Plains.

Red-necked Phalarope
Phalaropus lobatus

Robin-sized
Open Ocean; Inshore Waters;
Lakes and Ponds

Field Marks
7¾". A delicate shorebird with a needlelike black bill as long as its
head. From tundra breeding grounds, this species migrates both on
ocean and through interior West. Forms flocks of thousands that feed
on salt lakes and ponds, especially in fall. Swims lightly and spins,
picking plankton from water.
Breeding females (1) dark gray above, with golden stripes on back and
scapulars; throat white; deep rufous across upper breast and neck;
gray sides and flank streaks; rest of underparts white; white wing
stripe and sides of rump visible in flight. Breeding males similar, but
smaller and much duller, with little rufous. Winter adults medium
gray above, with broad white edges forming stripes down back; white
below; head white with blackish ear patch and nape. Juveniles (2)
brownish-black above, with golden edges forming back stripes; white
below, with pale buff collar.
Call a *tik*.

Similar Species
Red Phalarope larger and plumper; broader bill may be light in color.
Wilson's taller, with longer neck and bill, plain wing, white rump.

Range
Breeds in Alaska and n. Canada. Migrates along both coasts, more
common in West where it also appears inland. Winters at sea in
Southern Hemisphere.

Red Phalarope
Phalaropus fulicaria

Robin-sized
Open Ocean

Field Marks
8½". The most pelagic of shorebirds; away from tundra breeding grounds it is almost always on the open ocean. Irregularly comes to shore, and is even rarely seen inland, sometimes because of storms. Swims lightly on ocean, picking plankton from surface. Compact, thick build. Bill about length of head; broad and relatively thick. Breeding females (1) uniformly chestnut-red below, with white cheeks and black crown; back and scapulars black, with broad yellow-buff edges; rump chestnut; tail gray; wings gray with white stripe visible in flight; bill yellow-orange with black tip. Breeding male similar, but smaller and duller. Winter adults (2) clear medium gray above and on side of breast; white below and on face and crown; blackish ear patch and nape; bill dark, yellowish at base. Juvenile blackish above, with sharp golden feather edges; white below and on face, with buff-brown collar; bill mostly dark.
Call a high *kip*.

Similar Species
Red-necked Phalarope smaller, slimmer, with needlelike black bill; always has streaks on back. See Wilson's Phalarope and Sanderling.

Range
Breeds in Alaska and n. Canada. Migrates off Pacific and Atlantic coasts. Winters at sea in Southern Hemisphere.

Red Knot
Calidris canutus

Robin-sized
Beaches; Mudflats; Salt Marshes;
Freshwater Marshes

Field Marks
10½″. A large, compact sandpiper with straight, black bill, slightly
longer than head, and greenish legs. Gregarious, but most commonly
seen in small flocks. Mainly coastal, foraging on mudflats, beach
wrack, and near marshes; marches along water's edge, probing
surface.
Breeding adults (1) have rufous-chestnut underparts and face, with
variable white on lower belly through undertail; upperparts black,
with large rufous spots and white fringes; rump white with narrow
dark bars; tail gray; narrow white wing stripe. Winter-plumage adults
(2) medium gray above, minutely streaked with brown; breast, neck,
sides, and flanks speckled brown on pale grayish wash; rest of
underparts white. Juveniles have finely streaked breast and neck;
back and wing feathers finely fringed white and dark.

Similar Species
Dowitchers have bill much longer than head; longer legs. Nonbreeding
Black-bellied Plover larger, with short, thick bill and speckled
upperparts. Related sandpipers much smaller.

Range
Breeds locally in arctic Canada. Winters along Pacific Coast from
California south and Atlantic and Gulf coasts from New York.
Migrates along Atlantic and Pacific coasts and through Great Lakes
region.

Sanderling
Calidris alba

Robin-sized
Beaches; Mudflats

Field Marks
8″. A compact sandpiper with a straight black bill as long as its head, and black legs. The wave chaser, running up and down sandy beaches to feed within inches of the surf. Abundant on coasts.
Breeding adults (1) have upperparts, head, neck, and breast rufous with black speckling; fresh plumage has broad white fringes; belly and flanks white; bold white stripe in wing, dark tail and central rump visible in flight. Winter-plumage adults (2) immaculate white below and on underwing; pale gray above; face mostly white; upper wing with blackish leading edge and bold white median stripe. Juveniles white below; black back, scapulars, and crown, heavily spotted with white, create checkered appearance; white and buff face; nape and sides of breast buff with smudgy streaks; black wrist.
Flight call a snappy *kip*.

Similar Species
Red Phalarope pelagic; in winter has black ear mark; lacks black leading edge of wing. All other sandpipers are darker than winter Sanderling. See Baird's Sandpiper.

Range
Breeds in Alaska and arctic Canada. Winters along Pacific Coast from British Columbia south and on Atlantic and Gulf coasts from Massachusetts south. Also occurs inland on lakes and rivers.

Purple Sandpiper
Calidris maritima

Robin-sized
Rocky Shores

Field Marks
9″. Darkest sandpiper in the East. Breeds in the Far North; in our range, seen chiefly in winter along rocky coastlines. Forages for small crustaceans and mollusks among wave-lashed rocks and in tide pools. Often very tame.

Breeding adults chunky-looking; streaked overall with dark slate-gray; head and neck more finely streaked; legs yellow-orange; bill long, sturdy, slightly downcurved. Winter adults (1) paler, less streaked above; back and wings may show purple gloss; underparts streaked with gray-brown, except for white belly; white spots before eye visible at close range. Juveniles resemble winter adults, but may have buff and chestnut edgings on wings.

Gives twitters and a flight call, *prrt-prrt*.

Similar Species
Dunlin in winter plumage paler above and below, with black legs; flanks white or pale gray, not streaked.

Range
Breeds in n. Canada and Greenland. Winters on Atlantic Coast from Canada to Virginia, occasionally farther south. Rare visitor to Great Lakes.

Pectoral Sandpiper
Calidris melanotos

Robin-sized
Freshwater Marshes; Salt
Marshes; Mudflats; Ponds

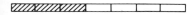

1

Field Marks
8¾". Large, chestier, and longer-necked than related small
sandpipers. Uses varied shore habitats, but prefers marshy, grassy
edges. Most often seen singly or in small flocks with other species. Bill
slightly longer than head, slightly downcurved, with pale base.
Breeding males have blackish-brown upperparts, with broad rufous-
buff fringes; breast and neck heavily streaked and mottled with dark
gray-brown, ending abruptly at boundary with white belly; head finely
streaked with brown; gray-brown upper wing has faint stripe; legs
yellowish. Breeding females similar but smaller. Adults in winter
similar but duller, with fairly plain brown back. Juveniles (1) similar
below, with finer streaks on buff breast and neck; black-brown above,
with broad rufous-buff and white fringes, which tend to form lines
on back.
Flight call a low, husky *chrrrit*.

Similar Species
Least Sandpiper similarly colored but much smaller; gives a shrill
kreet call. Baird's and White-rumped sandpipers slimmer, with longer
wings, shorter neck; breast streaks weak.

Range
Breeds in n. Canada. Migrates in fall along Pacific and Atlantic coasts,
inland in spring. Winters in South America.

Semipalmated Sandpiper
Calidris pusilla

Sparrow-sized
Mudflats; Beaches; Salt Marshes;
Freshwater Marshes

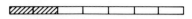

Field Marks
6¼″. A tiny sandpiper that specializes on mudflats, but also occurs on beaches and around marshes. Abundant on Atlantic Coast, scarce in West. Forages by quick picking and shallow probing. Black bill is length of head or shorter; relatively thick, especially at tip.
Breeding adults (1) have blackish-brown upperparts, with broad gray-buff fringes; white below, with fine dark streaks on breast and sides; upper wing gray-brown, with thin white stripe; tail and central rump dark; legs black. Winter adults medium to pale gray above, white below; minutely streaked with brown on upperparts; band across breast. Juveniles (2) black-brown above with broad rufous-buff and white fringes on back; white below, with breast washed pale buff and minutely streaked on sides.
Flight call a low, husky *churk*; also twitters in flock.

Similar Species
Western Sandpiper has chestnut fringes above in breeding and juvenal plumages; bill usually longer, thinner-tipped; call higher, thinner. See Least, White-rumped, and Baird's sandpipers, Dunlin, and Sanderling.

Range
Breeds in n. Canada and winters in South America, rarely s. Florida. Migrates along Atlantic Coast and inland, less commonly along Pacific Coast.

Least Sandpiper
Calidris minutilla

Sparrow-sized
Mudflats; Freshwater Marshes;
Salt Marshes; Beaches

Field Marks
6″. Our tiniest sandpiper, and the most widespread. Found in varied
fresh- and saltwater habitats, but prefers to have cover nearby.
Forages by quick picking and probing. Black bill is slightly shorter
than head and very fine, especially at tip.
Breeding adults have blackish-brown upperparts, with gray-buff to
rufous fringes; breast, neck, and sides heavily but finely streaked with
brown and washed with buff; belly white; upper wings gray-brown
with faint white stripe; tail and center of rump dark; legs yellowish.
Winter adults (2) dark grayish-brown above; breast and neck pale
gray-brown, with faint streaks; belly and flanks white. Juveniles (1)
black-brown above, with rufous and white fringes; white fringes form
converging lines on back; wing coverts fringed with buff.
Flight call shrill, rising *kreet*, with *ee* sound emphasized.

Similar Species
Semipalmated and Western sandpipers whiter on breast, with black
legs; paler and grayer above in winter; different calls. Other
sandpipers larger, have different calls; most have dark legs.

Range
Breeds from n. Alaska across n. Canada to Newfoundland. Winters on
coasts from Oregon and North Carolina south. Widespread inland
during migration and locally in winter.

White-rumped Sandpiper
Calidris fuscicollis

Sparrow-sized
Mudflats; Beaches; Salt Marshes;
Freshwater Marshes; Open
Country

Field Marks
7½". Easiest of the small sandpipers to identify, because of its white
rump, which is conspicuous in flight. Inhabits mudflats, beaches,
fresh- and saltwater marshes, and open country. At rest, wing tips
extend beyond tail—another useful field mark. Often mixes with large
flocks of assorted shorebirds; distinctive call will help identify it.
Breeding adults (1) spotted and streaked with dark brown above; ears
have some chestnut color, and back and wings have some bright buff
between the streaks; breast and flanks marked by narrow rows of fine
dark brown streaks; belly clear white. Winter adults dark above and
on upper breast, without distinct pattern of streaks; belly white;
eyebrows white, visible at close range. Juveniles (2) have spotted,
chestnut-fringed upperparts; white eyebrow; and incomplete buff V on
back. All plumages exhibit white rump in flight.
Call a thin *tseep* or *tzeet*.

Similar Species
Western Sandpiper slightly smaller; breeding adults much more
brightly marked with buff-gold. Winter Western and Semipalmated
sandpipers have black line on rump in flight; upper breast paler; wing
tips do not extend beyond tail.

Range
Breeds in n. Canada and winters in South America. Migrates along
Atlantic Coast in fall, inland in spring.

184

Western Sandpiper
Calidris mauri

Sparrow-sized
Mudflats; Beaches; Salt Marshes;
Freshwater Marshes

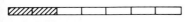

Field Marks
6½″. A tiny sandpiper that frequents mudflats, but also occurs on
beaches and around marshes. Common on Atlantic Coast, more
abundant on Pacific. Forages by quick probing. Black bill is as long as
head or slightly longer, with fine tip and usually a slight droop.
Breeding adults (2) have upperparts of mixed black-brown and
chestnut, with gray-buff fringes; ear patch and tinge on crown rust;
white below, with dark streaks on neck and breast becoming chevrons
on sides and flanks; upper wing gray-brown with thin white stripe; tail
and center of rump dark; legs black. Winter adults (1) medium to pale
gray above, mainly white below; minutely streaked with brown on
upperparts; band across breast. Juveniles black-brown above with
broad chestnut and white fringes on back and scapulars; white below,
with breast washed pale buff and minutely streaked on sides.
Flight call high, thin *jeet*. Also twitters in flock.

Similar Species
Semipalmated Sandpiper has buff above in breeding and juvenal
plumages; usually shorter, thicker-tipped bill; lower call. See Least,
White-rumped, and Baird's sandpipers, Dunlin, and Sanderling.

Range
Breeds in nw. Alaska. Winters along coasts from California and
Virginia south. Migrates along both coasts.

Buff-breasted Sandpiper
Tryngites subruficollis

Robin-sized
Open Country; Ponds

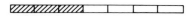

Field Marks
8¼″. A medium-sized sandpiper that breeds in the Far North; in migration, seen chiefly in open countryside or near ponds in grassy areas. Known for wing-raising behavior, performed mainly in courtship display on arctic breeding grounds, sometimes during migration.
Adults have rich buff underparts, darkest on breast and upper belly; above, have dark blackish-brown, scaly-looking pattern on back and wings; face buff; crown speckled with dark blackish-brown; legs and feet yellowish; bill rather short, slender and tapering toward tip. Juveniles (1) similar but somewhat paler below, especially on belly. All plumages show white wing linings in flight.
Gives a *tic* or *tic-tic-tic* and a low, soft, trilled *preet*.

Similar Species
Upland Sandpiper larger, with smaller head, thinner neck, and longer tail; underparts have brown bars and streaks. Juvenile Baird's Sandpiper has shorter black legs; wing tips extend beyond tail.

Range
Breeds in n. Alaska and arctic Canada; winters in South America. Migrates through interior, rarely along Atlantic Coast in fall.

Baird's Sandpiper
Calidris bairdii

Sparrow-sized
Open Country; Mudflats; Beaches;
Freshwater Marshes; Salt
Marshes

Field Marks
7½". A buffy sandpiper with a long, horizontal look, partly because
the wings extend past the tail tip. Usually in small flocks at wet
grassy edges. Walks quickly, but often picks for food deliberately.
Most common in mid-continent. Straight black bill as long as head.
Breeding adult (2) has blackish-brown upperparts, with extensive buff
fringes; breast, neck, and sides finely streaked with brown and
washed with pale buff; belly and flanks white; upper wing gray-brown
with faint stripe; tail and center of rump dark; legs black. Winter
adults gray-brown above; white below, with gray-buff, faintly
streaked breast. Juveniles (1) brown above with scaly look; white
below, with buff, faintly streaked breast, neck, and face.
Flight call low, rough *krrrit*.

Similar Species
White-rumped Sandpiper has white all across rump; different call;
more streaked and less buff. Sanderling stouter; wings do not extend
well past tail tip; bold wing stripe; breast rufous in breeding, white
in other plumages. See Buff-breasted and Pectoral sandpipers.

Range
Breeds in Alaska and arctic Canada; winters in South America.
Migrates mainly through interior, occasionally along Pacific and
Atlantic coasts in fall.

Solitary Sandpiper
Tringa solitaria

Robin-sized
Freshwater Marshes; Salt
Marshes; Ponds; Streams;
Mudflats; Northern Wooded
Swamps; Southern Wooded
Swamps

Field Marks
8½″. Truly a solitary sandpiper on most occasions. Usually at freshwater edges, especially vegetated shores, often away from other shorebirds.
Breeding adults (2) dark brown above, with numerous white flecks; head, neck, and breast finely streaked with brown; belly white; wings plain dark on both surfaces; rump and tail dark in center, barred with white on sides; narrow white eye-ring; legs greenish, delicate; dark bill short, straight, needlelike. Winter adults (1) plainer gray, with some white flecks on back and wings; eye-ring shows stronger against gray head. Juveniles like winter adults, but darker olive-brown above, with many buffy flecks; brownish-gray wash across breast and neck. Rather quiet, but gives a high-pitched *peet-weet*, often repeated.

Similar Species
Lesser Yellowlegs larger, has longer, yellow legs; white rump. Spotted Sandpiper has white on both wing surfaces; flies on shallow, stiff wingbeats; shorter legs, neck, and bill; pale bill; lacks light flecks above.

Range
Breeds from Alaska across Canada to Labrador and south to ne. Minnesota. Winters mostly south of United States, locally along Gulf Coast. Migrates throughout United States.

Spotted Sandpiper
Actitis macularia

Sparrow-sized
All freshwater habitats; Salt
Marshes; Mudflats; Beaches

Field Marks
7½". Usually seen singly, this small shorebird constantly bobs up and down while holding its head still. Style of flight—on shallow, stiff, buzzy wingbeats—is unique. Found in all freshwater habitats, including small streams. Also visits shores, preferring the border between rocks and beach.
Breeding adults (2) white below, thickly peppered with round blackish spots; olive-brown above, with blackish spots and bars; dark eyeline and white eyebrow; bill orange-pink with dark tip; wings brown with white stripes on both surfaces; tail brown with white barred edges.
Winter adults (1) plain olive-brown above; white below, with distinctive gray patch extending down side of breast; bill duller.
Juveniles like winter adults, but wing coverts barred with buff and brown, and rest of upperparts faintly tipped with buff.
Loud *peet-weet!* is often repeated.

Similar Species
Solitary Sandpiper has dark wings without white stripes; does not have stiff-winged flight; longer legs, neck, and needlelike dark bill; light flecks above; call higher-pitched.

Range
Breeds throughout most of n. United States and Canada. Winters along Pacific Coast from British Columbia south; along Atlantic and Gulf coasts from South Carolina; and inland through s. United States.

Common Ground-Dove
Columbina passerina

Sparrow-sized
Open Country; Thickets

Field Marks
6½". Smallest North American dove; sparrow-sized but plumper.
Found in a variety of open to mostly open habitats, including
developed areas. Head and breast distinctly scalloped; chestnut
primaries and wing linings conspicuous in flight; tail short, fan-shaped,
black. Feeds on ground, nodding its head as it walks. Usually quite
tame and easily approached.
Adults (1) grayish-brown, with head and breast scalloped; upperparts
browner, with black spots on wings; primaries and wing linings
chestnut; tail short, fan-shaped, and mostly black with small white
corners. Males have grayish crown, pinkish-gray breast. Females and
immatures paler and grayer.
A soft, repeated *woo-ooo*.

Similar Species
Only tiny dove in East. Inca Dove (in West) scalloped overall; tail long
with more white, less black edging.

Range
S. California east across southernmost states to Florida and north
along Atlantic Coast to North Carolina.

190

Mourning Dove
Zenaida macroura

Robin-sized
Forest habitats; Open Country;
Residential Areas; Urban Areas

Field Marks
12″. Our most widespread and common native dove. Found in open woodlands, many upland habitats, urban and suburban areas; common at feeders. Long, pointed, white-edged tail distinctive in flight. Wings whistle on takeoff. Gregarious. Usually feeds on ground, primarily on seeds.

Adults (1) brown above with conspicuously darker tail and wings, latter with dark spots; nape and sides of neck with purple sheen; underparts grayish with pink wash; tail long, pointed, edged with white; black spot on side of head; bill dark. Juveniles scaled and spotted on wings and breast, without pink wash on underparts; resemble adults by first winter.

A mournful cooing familiar to many: *who-ooh, who-who-who*. Second note rises sharply; others on same pitch. First two notes often not heard.

Similar Species
White-winged Dove has large white wing patches; rounded tail with white corners. Inca Dove smaller; plumage scalloped; shows chestnut primaries in flight; tail rounded. See Ringed Turtle-Dove and White-tipped Dove.

Range
Breeds from Alaska across Canada and south throughout entire United States. Some northern birds move south in winter.

191

Inca Dove
Columbina inca

Sparrow-sized
Texas: Open Country; Streamside
Forest

1

Field Marks
8¼". A small ground dove common in arid and semiarid habitats, urban and suburban areas. Smaller and slimmer than Mourning Dove; plumage distinctly scalloped. Chestnut primaries and wing linings conspicuous in flight. Often in flocks in winter.
Adults (1) pale gray, with conspicuous scalloping above and below; primaries and wing linings chestnut; tail long, rounded (at rest may look pointed), edged with white. Immatures similar to adults.
Call a monotonous, repetitious *coo-oo*, accent on second syllable; sometimes rendered as *whirl-pool*.

Similar Species
Mourning Dove larger; browner; tail pointed; lacks scalloping, chestnut primaries, and wing linings. Common Ground-Dove has shorter tail with less white, more black edging; scalloped on head and breast only. White-tipped Dove larger; lacks scalloping, chestnut primaries; usually in dense cover.

Range
Resident from se. California to south-central Texas and Central America.

Ringed Turtle-Dove
Streptopelia risoria

Robin-sized
Urban Areas; Residential Areas
and Parks

Field Marks
11″. An introduced species with wild populations in some southern
cities and towns. Domesticated for many centuries; today's wild
populations probably descended from escaped cage birds. Typically in
parks, gardens, and residential areas. Often perches on telephone
wires. Approximate size and shape of Mourning Dove, but much
paler.
Adults (1) pale tan overall, with head and underparts somewhat paler
than upperparts; narrow black collar around back of neck; bill dark;
tail rounded with white corners. Immatures similar to adults, but may
lack black neck collar.
Gives a soft, cooing *coo-curroooo*.

Similar Species
Mourning Dove much darker; tail pointed, edged with white; lacks
black neck collar.

Range
A domestic species, escaped and introduced usually in or around
cities: Los Angeles and s. California; Miami, Tampa, and s. Florida;
Arizona and Alabama.

White-tipped Dove
Leptotila verreauxi

Robin-sized
Southern Texas: Streamside
Forest

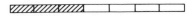

Field Marks
11½". A shy, plump inhabitant of streamside forested areas in
southernmost Texas; feeds mainly on the ground, but will also fly high
in canopy to take nuts and berries. When flushed, wings produce a
whistling noise. Formerly called White-fronted Dove.
Adults (1) basically brown above, with paler forehead and belly;
chestnut wing linings and white feathers at corners of tail show
plainly in flight.
Gives a long, low *cooo*; lower-pitched than calls of most other North
American pigeons and doves.

Similar Species
Common Ground-Dove much smaller, with darker underparts and
dark brown spotting on wings. Inca Dove much smaller, with scaly
pattern above and below. Mourning Dove has long, pointed tail;
underparts gray with rose-pink to buff wash. White-winged Dove
shows blackish wings with white patches in flight; at rest, has sandy
or brownish underparts and narrow white wing patch.

Range
Resident locally in s. Texas. More common in Central and South
America.

White-winged Dove
Zenaida asiatica

Robin-sized
Southern Texas: Brushy Open
Country; Streamside Forest;
Residential Areas and Parks

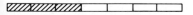

Field Marks
11½". A large, heavy dove of primarily arid habitats. Unmistakable in flight, with large white wing patches and white corners on tail. Approximate size of Mourning Dove, but heavier build, no pinkish cast to underparts. Gregarious; often with Mourning Doves where ranges overlap. Feeds mostly on seeds and fruit.
Adults (1) grayish-brown with nearly blackish wings and tail; crown and nape with purple sheen; large white wing patches conspicuous in flight, at rest look like white border on folded wing; tail rounded with large white corners conspicuous in flight; black streak on cheek. Immatures similar to adults, generally paler.
Cooing song variable; usually rendered as *Who-cooks-for-you?*

Similar Species
No other North American dove or pigeon has large white wing patches.

Range
Breeds from s. California to s. Texas. Winters mostly south of United States. Seen in small numbers in fall and winter on Pacific coast of California.

Rock Dove
Columba livia

Pigeon-sized
Urban Areas; Residential Areas
and Parks; Open Country;
Mudflats

Field Marks
12½″. A large, stocky pigeon, introduced from the Old World, now
common to abundant in cities, towns, and agricultural areas. Plumage
highly variable; wild form has dark head, gray body and wings, but
many color forms are seen, including nearly all-white, black, reddish-
brown, or speckled. Commonly roosts and nests on man-made
structures.
Adults of typical ancestral wild type (1) have dark, iridescent head;
paler gray body and wings with two dark wing bars; whitish rump;
gray tail with dark tip. Immatures resemble adults.
Soft cooing notes, sometimes rolled.

Similar Species
Larger and stockier than other doves or pigeons likely to be seen in
same habitat.

Range
From Alaska across s. Canada throughout United States to South
America.

White-crowned Pigeon
Columba leucocephàla

Pigeon-sized
Southern Florida: Mangroves;
Forest habitats

Field Marks
13½". A large, dark pigeon with white or pale crown. Enters our
range in southern Florida, where it inhabits mangroves and forested
areas. Most often observed perched in trees or flying overhead in
flocks. Takes insects and seeds as well as fruit.
Adults (1) appear all-black at a distance; crown white. Females
similar, with pale gray replacing white on crown. Immatures have
brownish crowns. In good light, all plumages show some green
iridescence about nape of neck.
Gives a loud, deep, owl-like *coo-coo-co-wooo* or *coo-cura-coo*.

Similar Species
Rock Dove lacks white crown, has two parallel dark bars on wings,
more evident iridescence on head and neck; generally not found in
mangrove areas.

Range
Breeds and winters in s. Florida and south.

Red-billed Pigeon
Columba flavirostris

Pigeon-sized
Southern Texas: Streamside
Forest

Field Marks
14½". A retiring, fairly uncommon pigeon that reaches our range only in dense streamside forests of southern Texas. Feeds in tops of trees, taking figs and other fruits, nuts, and seeds. Flies rapidly and directly on pointed wings.
Adults (1) appear entirely sooty gray from a distance; at close range in good light, dark maroon-brown of head, neck, and breast may be visible; bill short, with red at base and pale tip. Immatures resemble adults.
Call a rich, melodious cooing: *coooo, wup-cup-a-koo.*

Similar Species
Dark individual Rock Doves have dark bars on wings and white patch on rump; usually not all-dark.

Range
Local in s. Texas, chiefly lower Rio Grande Valley; winters mostly south of the United States.

Plain Chachalaca
Ortalis vetula

Crow-sized
Southern Texas: Streamside
Forest

Field Marks
22″. Unmistakable. A very vocal, long-tailed, chickenlike bird of streamside forests in southern Texas. Plain Chachalacas perch in bushes and trees, and can be somewhat secretive, but whole flocks sometimes join in noisy chorus in early morning. Often seen walking on ground.
Adults (1) plain olive-brown to gray-brown, darkest on upperparts; head small, with chickenlike bill; tail very long; wings appear short and rounded in flight; small patch of reddish skin sometimes visible on throat. Immatures similar to adults.
Male's call a deep, low *cha-cha-lac*, repeated often in rhythm; females and immatures have much higher calls. Birds in flocks sing at once but not in unison.

Range
Resident from s. Texas south to Central America. Introduced on islands off the coast of Georgia.

199

Greater Roadrunner
Geococcyx californianus

Crow-sized
Open Country

Field Marks
23″. Unmistakable. A large, usually shy, ground-dwelling cuckoo, typically seen running rapidly, neck outstretched, across roads. Seldom flies. Widespread and fairly common in arid and semiarid habitats through most of its range. Chases and preys upon insects, reptiles, and other small animals.

Adults (1) appear dark above with light streaking, buff below with dark streaks on breast; tail long, black with white edges; legs and neck long; sturdy bill; bare blue-and-red skin behind eye; may erect bushy crest. Immatures more reddish-brown above, without blue-and-red patch behind eye.

Song, lasting four to five seconds and often repeated, consists of six to eight down-slurred, dovelike coos, each slightly lower in pitch than preceding note.

Range
Local from n. California east to Arkansas and Louisiana, and south to Mexico.

Ring-necked Pheasant
Phasianus colchicus

Goose-sized
Open Country and Grasslands;
Freshwater Marshes; Residential
Areas

Field Marks
33″. A large, introduced game bird, widespread in grassy and brushy areas and woodland edges. Usually seen walking on ground, where it feeds primarily on grains, seeds, berries. When flushed, flies up with a loud whir of wings. Sociable; single males often accompanied by several females.

Males (1) unmistakable, with long, pointed tail; dark, glossy head with bare red facial patches; iridescent, multicolored body appears reddish-brown from a distance; usually has partial white neck ring. Females (2) smaller and duller; primarily buff, mottled with brown and black; tail long and pointed but shorter than male's. Both sexes have chickenlike bill; short, rounded wings. Immatures resemble adult females.

Explosive, hoarse *caaw-cák*, lasting about one second.

Similar Species
Female resembles Sharp-tailed Grouse, which has shorter tail with white edges and white undertail coverts.

Range
Resident in agricultural lands from Alberta, Washington, and n. California to New England. Local elsewhere.

Scaled Quail
Callipepla squamata

Robin-sized
Texas: Brushy Open Country

Field Marks
10″. Fairly common bird of arid scrub and grasslands; it feeds on the ground, primarily on insects and seeds. Combination of white- or buff-tipped crest and sandy-brown plumage with dark scaling distinctive. Usually escapes disturbance by running; seldom flies. Often in coveys, especially in winter.

Adult (1) basically sandy brown with dark scaling on blue-gray upper back, neck, and breast. Crest tipped white on male; buff-tipped and smaller on female. Some Texas males show chestnut belly patch. Immature resembles adult, but browner with mottled wings and duller scaling.

Two-noted, repeated *kip-kurrrr*; also a loud, strident whistle.

Similar Species
Northern Bobwhite darker, with black or dark brown face pattern, dark necklace; not found in arid areas.

Range
Resident from s. Arizona to sw. Kansas, south to Texas and central Mexico.

Gray Partridge
Perdix perdix

Pigeon-sized
Open Country and Grasslands

Field Marks
12½". An open-country, introduced game bird, widespread in cultivated and grassy fields. Plump, quail-like body. Rusty outer tail feathers conspicuous when flushed. Often forms coveys in fall and winter.

Adults (1) have rusty face and throat (less conspicuous in female); body brownish above with fine white streaks; breast gray, finely barred with brown; belly white, with conspicuous brown patch on male; flanks barred chestnut. Immatures similar; have yellow legs and feet instead of gray.

Calls include a loud, sharp *kee-ick*, soft clucking, and loud cackles.

Similar Species
Northern Bobwhite has dark outer tail feathers, black-and-white necklace.

Range
Resident from Washington and Oregon across s. Canada to Minnesota and Wisconsin, south to Utah and Nebraska. Scattered populations in north-central and ne. United States.

Black Francolin
Francolinus francolinus

Crow-sized
Louisiana: Open Country and
Grasslands

Field Marks
14″. An introduced, partridge-like game bird found only in southern
Louisiana; inhabits croplands and grassy areas, especially with plenty
of dense vegetation nearby for cover. Secretive; when startled by an
intruder, will run through cover, making observation sometimes
difficult.
Adult males (1, 2) have black head and breast, white cheek patch,
chestnut collar; upperparts mottled black and buff. Females mottled
brown and buff overall, with chestnut patch on back of neck; sandy
brown face has dark brown half-circle running from eye to chin.
Immatures similar to females. Legs and feet reddish-orange in all
plumages.
Call a buzzy, loud *click, kick-kirrick, kick, kikick*, or *chik, cheek-
cheek-cheek-chikick*; more insectlike than birdlike.

Similar Species
Northern Bobwhite much smaller, with black legs and feet and
distinctive whistled call. Prairie-chickens larger; ranges do not
overlap with Black Francolin's.

Range
Successfully introduced in sw. Louisiana.

Northern Bobwhite
Colinus virginianus

Robin-sized
Open Country and Grasslands;
Southeastern Pine Forest

Field Marks
9¾″. A small, plump, crested bird, named for its well-known call.
Primarily in East, where only native quail; introduced in parts of
West. Common in open woods, brushy areas, field edges; in coveys in
fall and winter. Feeds on ground, chiefly on seeds, grain, fruit. When
flushed, flies on noisy wings, then glides to cover.
Males (1) reddish-brown above and on sides, lighter below with
scaling and spotting; head blackish with white throat and broad, white
eyebrow; heavy black-and-white necklace; tail short, dark. Females
(2) duller, with buff throat and eyebrow. Immatures resemble adults.
Male's territorial call a clear, whistled, rising *bob-white* or *bob-bob-
white*. Both sexes give other whistled notes.

Similar Species
Gray Partridge has rusty outer tail feathers; lacks black-and-white
necklace.

Range
Resident from Kansas and Texas east to Massachusetts and Florida.

Willow Ptarmigan
Lagopus lagopus

Crow-sized
Boreal Forest

Field Marks
15″. A sturdy, plump grouse, widespread on arctic tundra and in boreal forest openings. Well-camouflaged year-round: white in winter, brown and white in summer. Very common some years, scarce in others. Feeds on buds, insects, fruit in summer; buds, twigs, catkins in winter. Feathered toes unique among North American grouse. Spring males (1) mostly white with reddish-brown hood; by midsummer (3) mostly reddish-brown with white wings and belly. Spring and summer females (4) brownish overall with dark barring; white wings often mostly concealed. Both sexes white in winter (2). Red eye-combs (dull on female, erect on male during courtship and display), white wings, and blackish outer tail feathers on both sexes in all plumages. Molting birds patchy white and brown.
Breeding male gives several loud, nasal calls, including a *kyow*; an accelerating roll; and short series of rapid *ba-cow* or *go-backa* notes.

Similar Species
Breeding male Rock Ptarmigan less reddish; winter male with black line through eye. Female Rock Ptarmigan difficult to distinguish; bill smaller, narrower.

Range
Alaska to Newfoundland south to central Canada. Wanders south to northernmost states.

Rock Ptarmigan
Lagopus mutus

Crow-sized
Boreal Forest

Field Marks
14". A widespread and common arctic grouse, similar to Willow
Ptarmigan but favoring higher, rockier, and more barren habitats.
Ranges farther north, but occasionally enters boreal forest.
Red eye-combs (dull on female, erect on courting and displaying
male), white wings, and blackish outer tail feathers show on both
sexes in all plumages. Spring males (3) mostly white with brown
barred hood; by midsummer, mostly brown (barred with black and
some white) with white wings and belly. Spring and summer females
(4) mostly brownish with dark barring; white wing often mostly
concealed. Both sexes white in winter (1, 2); male and some females
with black line from bill through eye. Molting birds patchy brown and
white. Immatures resemble adults by first fall.
Mechanical-sounding call is a toneless, nasal, rattling snore.

Similar Species
Breeding male Willow Ptarmigan redder; winter male without black
line through eye. Female Willow difficult to distinguish; bill heavier,
broader.

Range
Breeds in Alaska and arctic Canada, south to n. Quebec and
Newfoundland. Occasionally migrates farther south.

Sharp-tailed Grouse
Tympanuchus phasianellus

Crow-sized
Open Country and Grasslands;
Boreal Forest

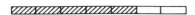

Field Marks
17″. A large, chickenlike bird, closely related and similar to prairie-chickens. Widespread but not conspicuous in grasslands, brush, or woodland edges. Usually on ground but flies strongly; sometimes perches in trees. Longish, pointed tail shows white edges and underside in flight. Like prairie-chickens, males gather on booming grounds in spring to display and to attract females.
Adults (1-3) mottled overall with pale brown and blackish, slightly darker above than below, spotted white on wings; belly whitish, spotted dark. Displaying male (3) shows purple neck sacs; has longer tail than female.
Muffled clucking or cackling and, in male display, dovelike cooing.

Similar Species
Female Ring-necked Pheasant lacks white outer tail feathers. Prairie-chickens barred, not spotted; have rounded, dark tails. See Ruffed Grouse.

Range
North-central Alaska to Ontario south to e. Oregon, ne. New Mexico, Wisconsin, and Michigan.

208

Spruce Grouse
Dendragapus canadensis

Crow-sized
Boreal Forest

Field Marks
16″. A widespread resident of northern spruce and pine forests and bog edges. Seen on ground or in trees. Usually exceedingly tame and easily approached; hence the nickname "fool hen."
Males (2) gray-brown above with black throat and breast; variable white tipping along sides, sometimes forming partial or complete breast band; red eye-combs conspicuous during courtship and display (3); short black tail, usually with chestnut tip. Females (1) variable; reddish- to grayish-brown; heavily barred; spotted white below on blackish background; tail dark with brown or buff tip; lacks red eye-combs. Immature resembles reddish female.
Breeding male gives very low-pitched booming; breeding female's call a series of rolled, clucking, and honking notes.

Similar Species
Ruffed Grouse crested; has longer and reddish-brown or gray tail with dark subterminal band and light tip; black feathers on sides of neck.

Range
Alaska to Nova Scotia, south to e. Oregon, Wisconsin, n. Michigan, n. New York, and n. New England. Also farther south in Rocky Mountains.

Greater Prairie-Chicken
Tympanuchus cupido

Crow-sized
Open Country and Grasslands

Field Marks
17″. A large prairie grouse, closely related and similar to Sharp-tailed Grouse. Once widespread and the common game bird of natural grasslands; now local and uncommon. Gathering on communal booming grounds in spring to display and attract females, males "dance" in hunched posture with tail raised, wings drooped, and long neck feathers erected to reveal yellow-orange neck sacs.
Adults (1–3) somewhat crested; brownish overall, barred above with dark brown and buff; lighter below with heavy, dark bars. Long, shaggy, black and whitish feathers on sides of neck. Short, fan-shaped tail dark in male (3), barred in female (1). Yellow eye-combs, long neck feathers, and yellow air sacs conspicuous in displaying males (2). Immature resembles adult by first winter.
Displaying male gives three-parted, low, resonant cooing; also cackles.

Similar Species
Lesser Prairie-Chicken slightly smaller with pale barring and reddish-purple neck sacs. Sharp-tailed Grouse scaled or spotted, not barred, with pointed, white-edged tail.

Range
Local in the Great Plains; scattered populations on Texas coast.

210

Lesser Prairie-Chicken
Tympanuchus pallidicinctus

Crow-sized
Open Country and Grasslands

Field Marks
16″. A chickenlike bird of arid grasslands and sandy plains, especially with scrub oaks; range restricted to small area in Southwest and Great Plains. Males perform group courtship displays, inflating air sacs on throat, fanning tail, raising feathers at back of neck, and bowing.

Adult males (3) buff, with dark brown barring overall; tail black; yellow combs over eye; when displaying (1), reveals red air sacs on neck. Female (2) similar, with barred tail; lacks eye combs and air sacs. Immatures similar to females.

Male gives bubbling, yodelling notes, in series of five; several series given in sequence. Both sexes cluck and cackle.

Similar Species
See Greater Prairie-Chicken. Sharp-tailed Grouse has long, pointed tail with white-tipped corners; frosty-looking wings in flight.

Range
Resident on southern Great Plains from e. New Mexico to Oklahoma and Kansas, and south to Texas Panhandle.

Ruffed Grouse
Bonasa umbellus

Crow-sized
Eastern Deciduous Forest; Boreal
Forest

Field Marks
17″. The widespread and common "partridge" of mixed woodlands.
Named for black feathers on sides of neck that male raises into a ruff.
Often unnoticed until explodes noisily into flight when flushed. In
spring, male perches on woodland drumming logs and beats wings in
an accelerating drum roll; sounds like a distant, muffled engine
starting up. Like Spruce Grouse, most conspicuous in summer, when
females are often seen with young.
Two distinct color phases: gray-phase adults (2) mottled grayish
above, lighter and barred below; tail gray, finely barred, with broad,
dark subterminal band and light tip. Reddish-brown replaces gray in
red phase (1). Both phases show small crest and black feathers on
sides of neck. Immatures resemble adults.
Male's drumming, produced with wings, functions as song. Also
various vocalizations, seldom heard.

Similar Species
Female Spruce Grouse lacks black neck feathers; has different tail
patterns. See Sharp-tailed Grouse.

Range
Resident from central Alaska across Canada to Labrador, south to n.
California, n. Utah, Colorado, Kentucky, and New Jersey, and in
Appalachians to n. Georgia.

212

Wild Turkey
Meleagris gallopavo

Very Large
Forest habitats; Open Country
and Grasslands

Field Marks
46″. Unmistakable. Largest North American game bird and ancestral
stock of barnyard turkey. Huge, with long legs, neck, and tail. Once
widespread, but reduced by clearing of woodlands. With management
and reintroduction, now increasing in many areas. Favors woodlands
and edges; feeds mostly on ground, primarily on seeds and nuts.
Roosts in trees. Sociable.
Adult males (1, 2) wholly dark with iridescent reddish and bronzy
cast; wings barred with white; head unfeathered, blue with red
wattles; breast with protruding "beard" feathers; tail and rump tipped
chestnut in East, whitish in West. Females smaller, duller; wattles
less conspicuous; often without beard. Immatures resemble female.
Male gives familiar gobbling; also *pit* notes, clucks, and others.

Range
Resident from se. California, Arizona, Texas, Southeast, and north to
New England. Small introduced populations elsewhere.

Turkey Vulture
Cathartes aura

Goose-sized
Open Country and Grasslands;
Forest habitats

Field Marks
27″. Large, dark, eaglelike. Widespread and common, conspicuous as
it soars over open country and woodlands. Flight profile distinctive:
broad, "fingered" wings held in shallow V as bird tilts from side to
side, seldom flapping. Head appears tiny relative to body. Roosts
communally. Feeds on carrion; several birds may gather at road kills,
often with Black Vultures where ranges overlap.
Perched or on ground, adults (2) uniformly dark brown overall
with small, unfeathered, red head; white-tipped bill; overhead (1),
pale gray primaries contrast with dark wing linings and body, giving
wings two-toned appearance; tail narrow and fairly long, extending
beyond feet. Immatures similar to adults, but with dark head and bill.

Similar Species
Black Vulture has white patches near wing tips, shorter tail, holds
wings flatter, mixes flaps with glides while soaring. Eagles larger and
soar on flat wings. See dark-phase Swainson's Hawk.

Range
Breeds from s. British Columbia across s. Canada to New England,
south to Mexico and Florida. Winters in the Southwest and from New
Jersey south.

Black Vulture
Coragyps atratus

Goose-sized
Open Country and Grasslands;
Forest habitats

Field Marks
25″. Large, dark, soaring bird of open country and populated areas.
Short tail and broad wings give chunky profile. Holds wings nearly
flat; alternates short glides with several quick flaps. Roosts
communally. Feeds on carrion, and often associates with Turkey
Vultures to feed.
Perched or on ground, adults (2) black above and below, with
unfeathered, gray head; light bill and legs. Overhead (1), wing tips
distinctly "fingered" and show conspicuous white patches. Immatures
similar to adults.

Similar Species
Turkey Vulture has longer tail and wings, latter appearing two-toned
overhead; holds wings in a distinctive V and seldom flaps while
soaring. Eagles much larger; lack distinctive flap-and-glide flight.
Immature Golden Eagle shows white wing patches, but also white at
base of tail.

Range
Resident from Pennsylvania south to s. Texas and Florida. Also in s.
Arizona.

Bald Eagle
Haliaeetus leucocephalus

Very Large
Lakes, Reservoirs, and Rivers;
Inshore Waters

Field Marks
40″. After California Condor, largest North American raptor. White head and tail of adult unmistakable. Long, broad wings held flat while soaring. Found mostly along seacoasts, large rivers, and lakes. Feeds primarily on fish, but occasionally takes carrion, small mammals, and waterfowl. Builds massive stick nest, used for several years.
Formerly more common in lower 48 states; now recovering in some areas from exposure to DDT and other toxins.
Adults (2, 3) brown above and below, with white head and tail; huge, yellow bill; yellow feet and eyes. Immatures and subadults (1) brown with varying amounts of white on underwings, body, and tail; bill dusky. Adult plumage usually acquired in fourth or fifth year.

Similar Species
Immatures and subadults similar to Golden Eagle; subadults have more sharply defined white patches on underwings and tail; smaller head and bill; shorter tail.

Range
Breeds in Alaska and across s. Canada, the Northwest, Great Lakes region, and locally along Pacific and Atlantic coasts. Winters mostly along rivers south of Canadian border.

Golden Eagle
Aquila chrysaetos

Very Large
Open Country

Field Marks
37″. A huge, soaring bird of hills and open country; rare in East;
widespread and fairly common in West. Close to size of Bald Eagle.
Often soars with huge, broad wings slightly uplifted. Feeds primarily
on rabbits and rodents, sometimes carrion.
Adults (1) brown above and below, with golden-buff wash on crown
and nape; bill dusky; eyes dark. Immatures and subadults (2, 3)
similar, but with white patches at base of primaries (more conspicuous
seen from below than above) and white base of tail. Adult plumage
usually acquired by fifth year.

Similar Species
Immature and subadult Bald Eagles usually show more white on
underparts; have larger head and bill; longer tail. Black Vulture
smaller, stockier, with white patches near wing tips, no white in tail;
flaps often while soaring.

Range
Breeds from nw. Alaska across Canada and south in the mountains to
Texas and Mexico. Isolated population in the Appalachians.
Sometimes wanders to mid-Atlantic states. Winters from British
Columbia to e. Canada and south into Mexico.

217

Harris' Hawk
Parabuteo unicinctus

Crow-sized
*Texas: Streamside Forest; Open
Country*

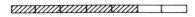

Field Marks
21″. A southern species that favors streamside forest, dry, open
scrub, and mesquite. Frequently soars close to ground; usually
perches low or on ground. Often seen in pairs or small groups.
Adults (1, 2) primarily dark brown, with chestnut on shoulders, wing
linings, and thighs; tail black with white base and tip. Immatures
duller and paler, especially on underparts; from below, body streaked
and tail banded; has blotched, chestnut shoulders. All plumages show
a white rump and base of tail in flight.
Most common call a loud hiss.

Similar Species
Only dark hawk with large areas of chestnut and white rump. Dark-
phase birds of other hawk species lack white rump, have broader tails.

Range
Resident in central Arizona, s. New Mexico, and s. Texas southward
to Mexico and Central and South America.

Crested Caracara
Polyborus plancus

Crow-sized
Texas, Louisiana, and Florida:
Open Country and Grasslands

Field Marks
23″. Unmistakable. An inhabitant of open country and brushlands.
Has long, sturdy legs; large, flat-topped head; large, eaglelike bill;
bare red face. Usually seen on ground, where feeds primarily on small
animals and carrion. Often feeds with vultures at road kills. Strong
flier with steady, shallow wingbeats; seldom soars.
Adults (2) have black crown with distinctive bushy crest; white face,
throat, and neck; dark body with light barring on upper back and
breast; white tail with dark tip and white wing patches conspicuous in
flight (1). Immatures similar, but browner and duller with light
spotting on wings, streaked breast.
Call a harsh *rak-rak-rak*.

Similar Species
Only raptor with large white wing patches. White-tailed Hawk paler,
with white underparts, gray crown and nape, rufous shoulders.

Range
Resident from s. Arizona, s. New Mexico, coastal Texas, sw.
Louisiana, and Florida south into tropics.

Snail Kite
Rostrhamus sociabilis

Crow-sized
Southern Florida: Freshwater
Marshes

Field Marks
17″. An uncommon, hawklike bird with a long, thin, strongly hooked
bill and broad white patch at base of tail. In our range, found only in
freshwater marshes of southern Florida. Bird consumes only one
species of snail—the apple snail, *Pomacea paludosa*—and is therefore
restricted to habitats supporting this mollusk. Flies gracefully on very
broad, rounded wings.
Adult males (1) dark slate-gray above and below, darker on head and
wing tips; tail shows narrow white margin at rest. Females (2) brown
above, streaked with brown and white below; face has white patch
over eye and white chin. Immatures similar to adult females. In all
plumages, note strongly downcurved bill and white patch at base
of tail.

Similar Species
Female and immature Northern Harrier have longer, narrower wings;
white patch on rump, not at base of tail; immature has cinnamon wing
linings.

Range
Resident and local in s. Florida; also in Central and South America.

Mississippi Kite
Ictinia mississippiensis

Crow-sized
Open Country and Grasslands;
Streamside Forest

Field Marks
14½". A gray hawk with pale head, usually found in open woods and grasslands, often near water. Distinctive in flight: graceful and buoyant, often glides. Falconlike, with longish tail and long, pointed, almost sickle-shaped wings. Perched, the long, crossed wing tips may look like a forked tail. Gregarious; may nest and roost in colonies. Feeds heavily on flying insects.
Adults (1) gray with obviously paler head; white patch, above, on inner wing conspicuous in flight; tail black; eyes red; feet yellow; bill small and dark. Immature slaty above; whitish below with heavy rusty streaking; tail black with three white bars and tip.

Similar Species
Black-shouldered Kite has black shoulders, mostly white tail. Falcons lack white wing patch above; have mustaches; flight feathers barred underneath (plain in immature Mississippi Kite); flight swifter, more direct.

Range
Breeds from Colorado and Arizona east to the Gulf Coast of Florida, north to Illinois and South Carolina. Winters in tropics.

Hook-billed Kite
Chondrohierax uncinatus

Crow-sized
Southern Texas: Streamside
Forest

Field Marks
16″. A rare visitor to southern Texas from tropical areas; has long oval
wings, narrow near the body and rounded at the tips, and fan-shaped
tail. Inhabits riverside forests and swampy areas, where it feeds on
various snails, extracting them from shells with prominent hooked
bill.
Adult males chiefly gray on head and back, with fine gray barring on
underparts; two wide white bands on underside of tail, and black-and-
white pattern of flight feathers distinctive in flight. Females (1)
similar to males, but gray replaced with dark sooty brown above,
reddish-brown below; rufous collar on neck. Immatures have black
forehead, crown, and nape; white collar; underparts less heavily
barred; upper wings brownish with some flecks of white. Black-phase
adults (extremely rare in Texas) almost wholly black, with some white
bars on wing tips and single white band on tail.

Similar Species
Gray Hawk similar to adult male, but has much smaller bill and lacks
black-and-white pattern on underside of wing tips. Compare flight
silhouettes of Broad-winged, Sharp-shinned, Cooper's, and Red-
shouldered hawks.

Range
Resident and local in s. Texas. Also in Central and South America.

Northern Goshawk
Accipiter gentilis

Crow-sized
Boreal Forest; Eastern Deciduous
Forest

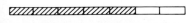

Field Marks
23″. Largest North American accipiter, widespread and uncommon in forested areas. Often the only winter accipiter in northern half of range. White eyebrows and fluffy white undertail coverts of adult conspicuous. Preys primarily on birds and mammals.
Sexes alike, but female larger. Adults (1) slaty above with dark crown and sides of head; white eyebrow; paler gray below, finely barred; fluffy white undertail coverts; longish tail banded, white-tipped; eyes red; feet yellow. Immatures (2) brown above, paler below with heavy brown streaks; light eyebrow (sometimes indistinct); yellow eyes and feet; tail has wavy blackish and gray bands; fine white lines between bands diagnostic.

Similar Species
Immature larger than Cooper's and Sharp-shinned hawks, which also lack fine white lines on tail. See dark-phase Gyrfalcon.

Range
Breeds from Alaska across Canada and south in the mountains to Mexico, South Dakota, the Great Lakes region, New England, and northern Appalachians. Winters in breeding range, occasionally farther south.

Osprey
Pandion haliaetus

Crow-sized
Inshore Waters; Lakes and
Reservoirs; Freshwater Marshes;
Salt Marshes

Field Marks
23″. A large, fish-eating hawk, usually seen along seacoasts, lakes, and
rivers. Widespread; fairly common along coasts, local inland. Easily
identified in flight by white underparts; long wings with conspicuous
bend at wrist; black wrist patches on underwing. Feeds almost
exclusively on fish, hovering over water, then diving, talons-first.
Like Bald Eagle, formerly much more common in lower 48 states; now
recovering in some areas.
Adult males (2) brown above with shaggy, mostly white head; brown
eye stripe; white underparts. Females (1, 3) very similar, with
scattered dark feathers across upper breast; in flight (1), tail and flight
feathers barred brown and white below, wing linings white with black
wrist patches. Immature similar; edged buff above, slightly buffy
below.
Typically a series of sharp whistles, the last often drawn-out.

Range
Breeds from nw. Alaska across Canada south to Baja California and
the Gulf Coast. Winters in s. United States and south to Argentina.

Gyrfalcon
Falco rusticolus

Crow-sized
Open Country; Beaches and
Coastal Dunes

Field Marks
23″. World's largest falcon; a widespread and rare breeding bird of
arctic habitats. Large, with a heavy build and broad-based wings; flies
on slower wingbeats than smaller falcons. Feeds primarily on birds,
especially ptarmigans. Nests on cliffs, occasionally in trees.
Plumage quite variable, ranging from entirely white with black bars
on back and wings (1) to gray above, whitish below with dark streaks
and bars; some birds mostly dark, streaked below (2); tail, barred or
unbarred, same color as back; eyes dark in all phases. Immatures
resemble adults.

Similar Species
Peregrine Falcon smaller, has dark hood and mustache; less uniformly
colored. See Northern Goshawk.

Range
Breeds from sw. Alaska across arctic tundra to Labrador. Winters
south to United States border, occasionally as far as California,
Colorado, Missouri, and Pennsylvania.

Black-shouldered Kite
Elanus caeruleus

Crow-sized
Southern Texas: Open Country

Field Marks
16″. A distinctive, graceful hawk of open grasslands with scattered trees and bushes. Adult unmistakable. Falconlike, with longish tail and long, pointed wings. Often hovers in near vertical position, hunting for rodents, reptiles, and large insects. Frequently seen on high, exposed perches.

Adults (1, 2) primarily pearl-gray above with conspicuous black shoulder patches, white tail, small bill; white below, primaries darker, with small black marks at wrists. Immature resembles adult, but with brown wash or streaking above and on breast; general pattern of adult recognizable.

Gives a rich, whistled *teew.*

Similar Species
See Mississippi and American Swallow-tailed kites. Male Northern Harrier has black on wing tips, not shoulder; in flight, wing linings white with black only on outer edges of primaries.

Range
Resident in s. California and s. Texas. May wander north to Oregon and east to Florida.

American Swallow-tailed Kite
Elanoides forficatus

Crow-sized
Southern Wooded Swamps

Field Marks
23″. A distinctive, easily recognized bird of prey with long, pointed wings and very long, forked tail. Inhabits wooded swamps of the Southeast and Gulf Coast; often seen near water. Takes flying insects on the wing, and captures other prey feetfirst, in Ospreylike fashion. Migrates in large flocks to wintering grounds in tropics.
Adults (1, 2) have white head, neck, breast, and belly; upperparts black; in flight, white wing linings contrast with black outer wing feathers; tail long, deeply forked. Immatures similar to adults, but show some streaking on underparts and head.
Gives a high-pitched *pee-pee-pee* or *eee-eee-eee*; also whistles.

Similar Species
Black-shouldered and Mississippi kites paler above, not so white below; lack long, deeply forked tail.

Range
Breeds locally on Gulf Coast from Louisiana to Florida and north to South Carolina. May wander north to Great Lakes and New England. Winters south of United States.

Short-tailed Hawk
Buteo brachyurus

Crow-sized
Southern Florida: Forest habitats;
Open Country

Field Marks

15½″. A stocky, rarely seen resident of forested areas, open country, and woodland edges in southern Florida. Soars in distinctive style, with outer primaries upturned. Two color phases; light-phase birds less common in our range.

Adult dark-phase birds black above with sooty black wing linings and pale gray undersides of primaries; tail gray with dark bands. Light-phase adults (1, 2) have white or pale wing linings and underparts. Immatures of both color phases resemble adults, but have more streaking on underparts and more bands in tail. Base of bill, legs, and feet yellow in all plumages.

Gives a high, squealed *kee-ah* when alarmed at nest.

Similar Species

Dark-phase birds unmistakable. Light-phase birds distinguished by style of soaring with upturned wing tips. Adult Red-shouldered and Broad-winged hawks have more bands on tail; underparts barred, not streaked.

Range

Breeds throughout most of Florida peninsula; withdraws to s. Florida in winter. Also in tropics.

White-tailed Hawk
Buteo albicaudatus

Crow-sized
Texas: Open Country and
Grasslands

Field Marks
23″. Dark above and pale below, this is a fairly common buteo of open
country and roadsides in southern Texas. Feeds on rabbits, lizards,
and small rodents. Often soars to great heights; flies with wings in a
shallow V. Perches close to ground on fence posts or bushes.
Adults (1, 2) gray above, white below, with rufous patches on
shoulders and narrow black band on white tail; in flight, show gray
wing linings and darker flight feathers. Immatures variable; mainly
dark above and below, with some blotchy white on breast and belly;
tail gray; in flight, wing linings appear darker than flight feathers.
Call a clear *ke-kee-kee* or *kee-ya, kee-yah*.

Similar Species
Dark phases of other buteos somewhat similar to immature, but
generally show some mottling, streaking, or banding on tail; fly with
wings more horizontal.

Range
Resident on coastal plain of Texas; also in Central and South America.

American Kestrel
Falco sparverius

Robin-sized
Open Country; Urban Areas;
Residential Areas and Parks

Field Marks
10½″. A small, widespread falcon, easily identified by its bright rusty back and tail. Feeding behavior distinctive: hovers while it looks for insects, small mammals and birds, reptiles; then stoops to take prey. Common in open country and populated areas. Quite vocal.
Adult males (2, 3) rusty above with striking blue-gray wings; cheeks white with two vertical black stripes; whitish below with rusty wash; back, wings, and flanks spotted or barred black; tail rusty with black subterminal band. Females (1, 4) rusty with darker barring on back and wings; whitish below with rusty streaks; head slightly duller than male's; tail rusty with dark bars. Immatures resemble adults but with heavier streaking.
A shrill, rapid series of *klee* notes, usually four or more.

Similar Species
Combination of small size and bright rusty back and tail distinguishes this from all other North American falcons and hawks.

Range
Alaska and Canada south to Mexico. Northern populations withdraw in winter.

Peregrine Falcon
Falco peregrinus

Crow-sized
Open Country; Freshwater
Marshes; Salt Marshes;. Urban
Areas

Field Marks
20″. Large, powerful, highly migratory. Flight swift and direct. In most of United States seen only on migration, usually along coast. Formerly widespread breeder across North America, but virtually eliminated by pesticides except in Far North and parts of West. Small numbers being reintroduced. Feeds primarily on birds.
Plumage varies regionally: arctic birds palest, northwestern birds darkest with heaviest pattern on underparts. Adults (1, 2) blue-gray to slaty above, with black cap and bold black mustache (narrowest in arctic birds); whitish to buff below, with barring usually restricted to breast and belly; tail long, banded; eyes dark. Immatures (3, 4) dark brown with buff bars above, heavily streaked below; mustache conspicuous.

Similar Species
Gyrfalcon larger, without prominent face pattern; more uniformly colored. Prairie Falcon has narrower mustache; shows black "wingpits" in flight. Merlin smaller, without prominent face pattern; tail has prominent bands.

Range
Breeds from n. Alaska and n. Canada south to the Northwest and to Mexico in the Rocky Mountains. Introduced pairs in the West and New England. Winters mainly in South America.

Merlin
Falco columbarius

Pigeon-sized
Open Country; Forest habitats;
Coastal Dunes

Field Marks
12″. A small, often low-flying falcon that feeds primarily on small birds (up to size of flickers). Swiftly overtakes prey; rarely stoops. Tail prominently banded. A northern breeder, usually near open country; highly migratory. Facial pattern less distinct than that of most falcons.

Plumage varies: Great Plains birds palest; Pacific Northwest birds darkest; others intermediate. Adult males blue-gray to slaty above; head brownish, streaked; whitish to buff below with dense, brown streaks; tail dark with prominent gray bands and white tip. Adult females (1) and immatures (2) resemble male but brown above, with buff tail bands.

Similar Species
Peregrine Falcon larger, with bold face pattern, less prominent tail pattern. Prairie Falcon larger; shows black "wingpits." American Kestrel has bright rusty back and tail, and prominent face pattern; often hovers.

Range
Breeds from Alaska across Canada and south to n. United States. Winters from southern parts of breeding range to South America.

Prairie Falcon
Falco mexicanus

Crow-sized
Open Country and Grasslands

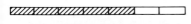

Field Marks
19½". A widespread falcon; most common in the West, in open country, especially near mountains or cliffs, but moves considerably eastward in winter. Unmistakable, with conspicuous black "wingpits" in flight. Feeds primarily on birds and small mammals.
Adults (1, 2) primarily brown above with narrow, dark mustache and dark eye patch, whitish below with brown streaks; underside of wings show black wingpits in flight; eyes dark. Immature similar to adult but darker, with heavier streaking.

Similar Species
All other falcons lack black wingpits. Peregrine Falcon has much bolder face pattern.

Range
Breeds from British Columbia to North Dakota, south to s. California and n. Texas. Winters in breeding range and east to the Mississippi River. May wander farther east.

233

Swainson's Hawk
Buteo swainsoni

Crow-sized
Open Country and Grasslands

Field Marks

21″. A common, medium-sized hawk of open western lands. In flight, shows profile like that of Turkey Vulture, with wings frequently held in shallow V. Highly migratory; often seen in large flocks on spring and fall flights. Feeds mostly on crickets and grasshoppers.

Plumage highly variable in both sexes. Typical light-phase adults (4) brown above with reddish hood; mostly white chin, belly, and wing linings that contrast with darker flight feathers; tail indistinctly barred with dark band near tip. Typical dark-phase adults all-brown with little contrast on underwing; tail as in light phase. Intermediate phases (1) often show hood. Dark-phase immatures similar to adults; light-phase immatures (2, 3) dark brown above with variable white markings; buffy white below, often with partial dark hood or breast collar and spots on belly.

A descending scream, *keeeer*, trailing off at end.

Similar Species

Red-tailed Hawk has wing linings same color or darker than light flight feathers, usually has rusty tail. See Rough-legged Hawk.

Range

Breeds from s. Alaska to Mexico, mainly in the West and on the Great Plains. May wander east to Great Lakes and Atlantic Coast. Winters in South America.

234

Northern Harrier
Circus cyaneus

Crow-sized
Freshwater Marshes; Salt
Marshes; Open Country and
Grasslands

Field Marks
22″. A widespread and common open-country hawk that quarters over ground, hunting primarily for small rodents. Easily identified in flight by white rump, long tail, and behavior: flaps occasionally, but mostly glides with long, narrow wings held up in shallow V. Except on migration, usually flies low. Formerly called the Marsh Hawk. Adult males (1, 2) gray above with black wing tips, white rump; white below. Adult females brown above with white rump; buff below with brown streaks. Immatures (3, 4, and inset) resemble females, but rich cinnamon below.
Gives a rapid *ke-ke-ke* at nest, silent elsewhere.

Similar Species
In flight, combination of white rump, narrow wings, and behavior unmistakable. In perched birds, long tail and owl-like face are distinctive.

Range
Breeds across Canada and south to s. United States and Mexico. Winters from s. Canada to South America.

Rough-legged Hawk
Buteo lagopus

Crow-sized
Open Country and Grasslands;
Freshwater Marshes; Salt
Marshes

Field Marks
22″. A large hawk of open country that breeds in Far North; winters
across much of United States. Highly variable, with light and dark
color phases. When hunting, often quarters over ground, harrierlike,
or hovers. Feeds primarily on rodents.
Adults and immatures of both sexes are highly variable; from below,
all birds show flight feathers and white, often banded, tail with dark
band near tip. Light-phase birds (1, right) most common; brown to
dusky above with pale outer wing patches and pale base of tail visible
in flight (inset); head and underparts paler, streaked and mottled
brown; black wrist marks; belly often blackish. Extreme dark-phase
birds (1, left) all dark except base of tail and undersides of flight
feathers. All birds have legs feathered to toes.

Similar Species
See Red-tailed Hawk, Northern Harrier, and Golden Eagle.

Range
Breeds in Alaska and n. Canada. Winters from s. Canada to
California, New Mexico, and Virginia.

Red-tailed Hawk
Buteo jamaicensis

Crow-sized
Open Country; Forest habitats

Field Marks
22″. Most common large buteo in North America, and one of most widespread. Habitat varies from open country to open woodland; rarely in dense forest. Large with typical buteo shape: wings broad and rounded, tail broad and fanned. Plumage highly variable. Often soars.

Darkest birds (1), found mostly in West, wholly dark brown, variably marked rusty with indistinctly banded or mottled reddish to whitish tail. Palest birds (Great Plains) pale brown above, nearly white below, with indistinctly marked, rust-tinged tail. Typical adults (2) brown above and on head; white below with variable dark belly band; rusty tail; dark leading inner edge on underwing. Immature variable, difficult to identify; usually brown above, white below with heavy spots and streaks; tail gray-brown, indistinctly banded. A loud, harsh scream, *keeeyerr*, trailing off and dropping at end.

Similar Species
Swainson's Hawk has white wing linings. See Rough-legged and Red-shouldered hawks.

Range
Breeds from s. Alaska across Canada and south throughout United States. Winters from s. Canada south to Central America.

Red-shouldered Hawk
Buteo lineatus

Crow-sized
Northern Wooded Swamps;
Southern Wooded Swamps;
Eastern Deciduous Forest

Field Marks
19″. A relatively long-winged, long-tailed buteo. All ages usually show characteristic pale patch at base of primaries in flight. Occurs mostly in woodland habitats, often near water; takes a variety of prey, including small mammals, snakes, and frogs.
Plumage variable in both sexes. Typical eastern adults (inset) brown above; wings darker with white mottling and rufous shoulders; barred reddish-brown and white below with reddish-brown wing linings; tail dark with prominent, narrow white bands. Florida residents (2) paler. Immatures (1) brown above; buff to white below with variable dark streaking; wing linings often reddish.
Up to four short, down-slurred *kee-yah* notes, at a distance similar to Blue Jay calls; also a longer *keeaarrr*, less harsh than Red-tailed's.

Similar Species
From below, Red-shouldered's pale wing patches usually distinctive in all plumages. Broad-winged Hawk chunkier, has paler underwing with dark outline, more uniformly brown upperparts. See larger Red-tailed Hawk.

Range
Breeds and winters in s. California and from the Great Plains, Quebec, and New England·south to Gulf Coast.

Broad-winged Hawk
Buteo platypterus

Crow-sized
Eastern Deciduous Forest

Field Marks
16″. A small, chunky hawk that nests primarily in eastern mixed forests; highly migratory. Well known for spring and fall flights, when thousands can be seen in a day along favored routes. Flight profile shows broad, almost pointed wings and short, fan-shaped tail. Takes a variety of small prey.

Adults (1) uniform brown above; white with rusty brown bars, densest on breast, below; tail dark with bold white bands and narrow light tip; in flight shows whitish underwing with dark outline.

Immatures (2) brown above with dark whisker mark; whitish below with dark streaks; underwing similar to adult's; tail less distinctly banded.

Two-noted, high-pitched whistle; first note short, second long; sounds like a squeaky brake.

Similar Species
Cooper's Hawk and Northern Goshawk have shorter wings, longer tails. Immature Red-shouldered Hawk darker below; wings longer, more rounded, with pale "windows." Other hawks in range larger. See Short-tailed Hawk.

Range
Breeds across s. Canada south to e. Texas and Florida. Winters in s. Florida, sparsely in California; also in Central and South America.

Sharp-shinned Hawk
Accipiter striatus

Pigeon-sized
Eastern Deciduous Forest; Boreal
Forest; Open Country; Coastal
Dunes

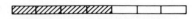

Field Marks
14″. Smallest and most common North American accipiter, widespread
in woodland habitats. Shows typical accipiter shape (rounded wings,
long tail) and flight (quick flaps interspersed with glides). Feeds
primarily on birds. Often migrates southward along coast, or along
mountain ridges.
Sexes similar; female larger. Adults (1) slaty above and on cap; barred
rusty and white below; undertail coverts white; tail square or slightly
notched with straight bands, narrow white tip; eyes red. Immatures
(2) brown above; whitish with reddish or brown streaks below; tail
like adult's; eyes yellow.

Similar Species
Easily confused with Cooper's Hawk, which is larger, has more
rounded tail with broader white tip. Adult Cooper's shows contrasting
darker cap; immature streaked darker below, often with unstreaked
belly. See Northern Goshawk, which is larger.

Range
Breeds from Alaska across Canada south to s. United States. Winters
from n. United States south to Central America.

Cooper's Hawk
Accipiter cooperii

Crow-sized
Eastern Deciduous Forest; Open
Country; Coastal Dunes

Field Marks
16″. A widespread, medium-sized accipiter, more common in western woodlands than in East. Closely resembles Sharp-shinned Hawk. Feeds primarily on birds and small mammals.
Sexes similar; female larger. Adults (2) slaty above with darker cap; barred rusty and white below; undertail coverts white; tail rounded with straight bands and broad white tip; eyes red. Immatures (1) brown above, whitish with dark brown streaks below; belly often unstreaked; tail like adult's; eyes yellow.

Similar Species
Adult easily confused with Sharp-shinned Hawk, which is smaller, has square of slightly notched tail with narrower white tip; adult lacks contrasting cap. Immature Cooper's confused with immature Sharp-shinned and with immature Northern Goshawk; Sharp-shinned square-tailed with more reddish streaking (including belly) below; Northern Goshawk heavily streaked below (including on belly and undertail coverts); has wavy tail bands separated by fine white lines.

Range
Breeds across s. Canada and throughout United States. Winters in central and southern states south to Central America.

Ferruginous Pygmy-Owl
Glaucidium brasilianum

Sparrow-sized
Southern Texas: Streamside
Forest

Field Marks
6¾". A small owl of arid and streamside habitats, rare and local in the
United States. Long-tailed; lacks ear tufts. Active by day as well as
night. Short, quick, shrikelike flight distinctive. Often flicks tail.
Easily imitated call often brings a response.
Color varies; sexes similar. Adults (1, 2) rusty to grayish-brown
above, with white or buff spots on wings; fine white streaks on crown;
black eyespots on nape; white with brown or rufous streaks below;
long tail rusty, with dark bars; eyes yellow. Immature resembles
adult, but lacks crown streaks.
Call is long series of *took* or *poop* notes, usually two notes per second
or faster.

Similar Species
Elf Owl smaller, short-tailed, without black eyespots.

Range
Local in s. Texas; also south-central Arizona. More common south of
United States.

Elf Owl
Micrathene whitneyi

Sparrow-sized
Southern Texas: Streamside
Forest; Brushy Open Country

Field Marks
5¾". Smallest North American owl. A nocturnal inhabitant of
streamside forests, canyons, foothills, and deserts with large cacti.
Feeds almost exclusively on large insects. Nests in old woodpecker
holes.
Adults (1, 2) grayish-brown above with conspicuous white shoulder
spots; whitish below with broad, blurry, brown streaks; eyes yellow,
eyebrows white; no ear tufts; tail short. Immature similar.
Calls include rapid series of short notes, dropping at end, and
soft barks.

Similar Species
Ferruginous Pygmy-Owl has longer tail, dark eyespots on nape.
Eastern Screech-Owl larger, "eared." See Northern Saw-whet Owl.

Range
Breeds in se. California (extremely rare), s. Arizona, s. New Mexico,
and s. Texas. Winters in Mexico.

Northern Saw-whet Owl
Aegolius acadicus

Robin-sized
Forest habitats

Field Marks
8″. Smallest owl in most of East. A widespread, fairly common inhabitant of dense coniferous forests and mixed woods, tamarack bogs, and thickets. Like closely related Boreal Owl, small, short-tailed, and lacks ear tufts, has flat-topped head. Strictly nocturnal and difficult to locate, but once found, tame and easily approached. Feeds primarily on insects.
Adults (2) brown above, with distinct fine white streaking on forehead and crown, large white spots on wings; white below with brown streaks; face whitish; bill dark; eyes yellow. Juvenile (1) distinctive: brown above, buff or rusty below, with broad white V on forehead; resembles adult by first winter.
Typical call is an easily imitated, deliberate series of *took* notes, often unevenly spaced.

Similar Species
Boreal Owl has black face border; spotted (not streaked) crown; yellowish bill. Eastern Screech-Owl lacks white face, has ear tufts. See Elf Owl.

Range
Resident from se. Alaska across Canada south to California and New Mexico in the West, Pennsylvania in the East. Winters south to Mexico and North Carolina.

Boreal Owl
Aegolius funereus

Robin-sized
Boreal Forest

1

Field Marks
10″. A tame but very inconspicuous denizen of dense forests in the Far North. Small; short-tailed; lacks ear tufts; has flat-topped head. Similar and closely related to slightly smaller Northern Saw-whet Owl. Feeds primarily on mice and insects.
Adults (1) brown above with contrasting fine white spots on crown and forehead, larger white spots on wings; white below with brown streaks; whitish face outlined with black; bill yellowish; eyes yellow. Juvenile brown, with whitish eyebrows; resembles adult by first winter.
Rapid series of nine to ten short *hoop* notes, rising slightly.

Similar Species
Northern Saw-whet Owl without black face border; has streaked (not spotted) crown, dark bill. Eastern Screech-Owl lacks white face; has ear tufts. Northern Hawk-Owl much larger than Boreal; barred below; long tail.

Range
Breeds from Alaska across Canada to s. Labrador, south to British Columbia and n. New England. Sometimes wanders farther south in winter.

Eastern Screech-Owl
Otus asio

Robin-sized
Eastern Deciduous Forest;
Southeastern Pine Forest; Open
Country; Residential Areas and
Parks

Field Marks
8½". Smallest owl with ear tufts in East. Has two color phases—red
and gray. Widespread and common in wooded areas, including
backyards. Strictly nocturnal. Often responds readily to imitation
of its call.
All plumages have small ear tufts, yellow eyes; bill usually light.
Adult plumage varies from rusty (1, left) to gray-brown (1, right)
above; paler face edged in black; white spots on wings; tail like back,
but indistinctly banded; underparts rusty to pale gray with dark
streaks and fine barring. Very young birds resemble adults, but
barred (not streaked), with smaller ear tufts.
Long, monotonous, rolled whistle; also a quavering, dropping whinny.

Similar Species
Voice diagnostic. Other eastern owls much larger or lack ear tufts.

Range
Resident from Manitoba, Ontario, and New England south to Texas
and Florida.

Northern Hawk-Owl
Surnia ulula

Pigeon-sized
Boreal Forest; Open Country

Field Marks
16″. A large, diurnal owl, widespread in open areas and forest edges of the Far North. Falconlike, with long tail, pointed wings, and swift flight, but with larger head. Usually flies low. Sits on exposed perches, often jerking tail up and down. Typically tame and easily approached.

Adults (1, 2) brownish above with white spots; heavily barred with brown and white below; whitish face outlined with black; lacks ear tufts. Immature resembles adult, but with less spotting above; barring below more reddish-brown; tail with broader white tip.

Similar Species
Combination of falconlike profile, diurnal habits, heavy barring below, and whitish face with black border makes this owl unmistakable.

Range
Resident from Alaska south to British Columbia and east across Canada. Winters south to northernmost United States.

Long-eared Owl
Asio otus

Crow-sized
Forest habitats; Groves

Field Marks
15″. A slender, medium-sized owl, widespread in coniferous and mixed forests of North America, but secretive and rarely seen. Roosts during day in dense cover, where well camouflaged. Long, dark ear tufts, closer together than on most other owls, and buff face are distinctive. Long wings and tail; flight butterflylike, with rapid upstroke of wings.

Adults (1, 2) mottled gray-brown above, similarly gray-brown below with dark streaks and white spots; ear tufts long, dark; face orange to buff (palest on western birds); eyes are yellow-orange. Immature resembles adult by first winter.

Varied calls include a series of drawn-out hoots, about one every two seconds (higher-pitched and slower than call of Great Gray Owl); also soft barks and drawn-out wails.

Similar Species
Great Horned Owl much larger with white throat; barred (rather than streaked) below; ear tufts farther apart. Eastern Screech-Owl much smaller. See Short-eared Owl.

Range
Breeds from north-central Canada east to Quebec, south to California, Texas, and Virginia; moves farther south in winter.

Great Horned Owl
Bubo virginianus

Crow-sized
Forest habitats; Open Country;
Residential Areas

Field Marks
22″. Large and powerful; widely distributed across North America.
Habitats vary from woodlands to open country, urban parks to semi-deserts. Combination of large size, prominent ear tufts set far apart,
bright yellow eyes, and white throat distinctive. Primarily nocturnal.
Can take prey as large as porcupine.

Plumage varies from very dark, in birds of the Pacific Northwest, to
very pale in Arctic. Typical adults (1, 2) mottled gray-brown above;
buff below, barred dark brown; ear tufts on head and white throat
distinctive; face tawny. Immature similar to adult but more rufous;
ear tufts smaller; white on throat less extensive.

Low, hooting *who-who-whowhowho-whooo-whooo*; sometimes
abbreviated to include only last three to four notes.

Similar Species
Long-eared Owl smaller; lacks white throat; streaked (not barred)
below; ear tufts closer together; richer buff. Other medium-sized to
large owls in range lack ear tufts.

Range
Resident from Alaska across Canada, south throughout United States
to South America.

Great Gray Owl
Strix nebulosa

Goose-sized
Boreal Forest

Field Marks
27". Largest North American owl. Favors dense coniferous and mixed forests; also second growth, especially near water in winter and migration. Face pattern, large rounded head without ear tufts, and long tail distinctive. Like Snowy Owl, irrupts southward in some years.

Adults (1, 2) brownish-gray above, streaked with brown and gray below; large face has concentric gray and brown circles, white crescents on throat; eyes yellow, appear small; bill light. Immature resembles adult.

Typically a series of up to 12 low *hoo* notes, given over eight to nine seconds, dropping slightly in pitch and speed.

Similar Species
Barred Owl much smaller, has dark eyes, more clearly barred or spotted plumage; lacks white throat crescents.

Range
Breeds from central Alaska and central Canada south to central California, Idaho, and northern Great Lakes region. Winters south to northernmost United States.

Barred Owl
Strix varia

Crow-sized
Forest habitats; Northern Wooded
Swamps; Southern Wooded
Swamps

Field Marks
21″. A large owl without ear tufts; very vocal, with a familiar hoot.
Found in wooded swamps of the North and South, as well as in other
forested areas. Feeds mainly at night, taking crayfish, frogs, and
small rodents. Range in West is expanding.
Adults (1, 2) primarily gray and brown, with white bars and edgings
above; underparts buff with black vertical streaks; face gray-brown;
collar barred horizontally, contrasting with streaked breast. Eyes
brown. Some individuals may be very pale.
Typical call a nine-noted hoot: *Who-cooks-for-you, who-cooks-for-
you-all?*

Similar Species
Great Gray Owl much larger, with yellow eyes and "bow tie" on
throat.

Range
E. British Columbia and the northwestern states across s. Canada,
south through e. United States to Gulf Coast.

Burrowing Owl
Athene cunicularia

Robin-sized
Open Country and Grasslands

Field Marks
9½". Unmistakable. A small, mostly diurnal owl of open country, including golf courses, airports, and vacant lots. Typically seen standing on ground on distinctive long legs; often bobs up and down. Rounded head without ear tufts; short tail. Nests (often colonially) in abandoned rodent burrows. Eats primarily insects and rodents. When flushed or startled, gives a loud chatter.

Adults (1, 2) primarily sandy brown above, with back, wings, and crown spotted white; whitish below with brown barring; throat and upper breast white, separated by dark collar; eyebrows white; eyes yellow. Juvenile distinctive, with no barring below; resembles adult by first winter.

Varied calls include shrieks, whistles, laughs; also a monotone *cook-coooo*, dovelike but higher-pitched.

Range
Breeds from British Columbia east to central Manitoba, and south to California and central Texas. Also resident in Florida. Northern birds move south in winter.

Short-eared Owl
Asio flammeus

Crow-sized
Open Country; Freshwater
Marshes; Salt Marshes

Field Marks
15″. One of our most commonly seen owls, occurring almost
exclusively in open country, feeding primarily at dawn and dusk.
In air, buff patches and black wrists on wings, and bounding,
butterflylike flight with quick upstrokes distinctive. When perched
low or on ground, looks relatively small-headed for an owl; ear tufts
usually inconspicuous.
Plumage varies from gray to brown. Typical adult (2) mottled brown
and buff above with white spots; tawny below with heavy dark
streaks on breast, finer streaks on belly; upper wing shows
conspicuous buff patches near tips, black wrists and tips; underwing
mostly pale with black wrists and tips; face buff or whitish; eyes
yellow. Immature (1) darker; resembles adult by first winter.
Harsh, nasal barks and squeals; does not hoot like most larger owls.

Similar Species
Long-eared Owl shows similar wing pattern in flight, but tip of
underwing appears pale, not blackish. See Barred Owl, Common
Barn-Owl.

Range
Breeds from Alaska across arctic Canada, and locally south to central
California, Kansas, Ohio, and New Jersey. Winters south throughout
most of United States.

253

Common Barn-Owl
Tyto alba

Crow-sized
All forest habitats except Boreal
Forest; All open, urban, and
residential habitats

Field Marks
16″. A slender owl found primarily in open to semi-open habitats, and
in cities and towns. Unmistakable: white, heart-shaped face unique
among North American owls. Has whitish underparts; long, feathered
legs; no ear tufts. Mostly nocturnal. Nests in hollow trees, caves, and
abandoned buildings.
Adults (1) golden buff above, conspicuously spotted black and white
on crown, back, and wing coverts; underparts usually whitish
(sometimes golden buff across breast), sparsely spotted black; heart-
shaped face white with tan border; eyes dark; bill light. Immature
resembles adult.
Gives an almost blood-curdling hissing or raspy scream.

Similar Species
Immature Snowy Owl lacks heart-shaped face; barred above in black
and white, below in black and buff. Adult Snowy nearly all-white.

Range
Breeds from British Columbia across n. United States to s. New
England and south throughout United States. Northern birds may
move south in winter.

Snowy Owl
Nyctea scandiaca

Crow-sized
Open Country and Grasslands;
Beaches and Coastal Dunes; Salt
Marshes

Field Marks
23″. Unmistakable. A large white owl of the Arctic, with striking
yellow eyes; resident in tundra and open country. Large, rounded
head without ear tufts; heavy build. Active by day as well as night.
Flight strong and direct; often glides low over ground. Nests in Far
North, where it feeds on lemmings and other small rodents; in
irruption years, when prey is scarce, invades southern Canada and
United States.
Adults (1) nearly all-white or with scattered dark bars above and
below; female with more barring than male. Immature (2) resembles
adult but more heavily barred and spotted.

Similar Species
See Common Barn-Owl.

Range
Breeds in n. Alaska and n. Canada. Winters south to northernmost
United States. Occasionally farther south in years of periodic
irruptions.

Lesser Nighthawk
Chordeiles acutipennis

Robin-sized
Southern Texas: Open Country
and Grasslands

Field Marks
9″. Closely resembles slightly larger Common Nighthawk. Found
primarily in dry habitats; flies low over ground, feeding on insects.
Most active at night, dawn, and dusk. Like other nightjars, has very
small bill, wide, gaping mouth; well-camouflaged during day when
resting horizontally on ground or a tree limb.
Adult male dark with gray, brown, and whitish mottling above; paler
below with dark barring; throat white; white band across primaries
conspicuous in flight; tail notched, with white subterminal band. Adult
female (1) similar to male but browner; throat buff; wing band
smaller, buff; tail lacks white band. Immatures similar to adults; may
lack wing band.
Call a froglike trill; also a low *chuck*.

Similar Species
Common Nighthawk's wing band always white, never buff; located
farther from wing tip. Other nightjars in range have rounded wings.

Range
Breeds from north-central California to s. Texas. Winters in tropics of
Central and South America.

Chuck-will's-widow
Caprimulgus carolinensis

Pigeon-sized
Forest habitats

Field Marks
12″. A large nightjar with reddish-brown plumage. Common locally in open woodlands and forest edges throughout most of the Southeast, especially near farmlands, streams, and rivers. Nests in leaf litter on forest floor. Sleeps by day; hunts at night, flying on rounded wings to take moths and other flying insects.

Adult (1) mottled reddish-brown above and below, with some brown on throat. Males and females almost indistinguishable from a distance; males have small white throat patch and some white in outer tail feathers, visible in flight; females have buff throat patch and no white in tail. Immatures resemble adults. All plumages show rounded wings in flight.

Call a loud, repeated series of four syllables: *chuck-will's-wi-dow*; first note low, last three whistled. Also gives a froggy *chuck* in flight.

Similar Species
Whip-poor-will smaller and much grayer overall. Common Poorwill much smaller; has white outer tail corners in flight. Common Pauraque has white bar on flight feathers; tail longer, with white outer edges.

Range
Breeds from e. Kansas east to New Jersey and s. New England, south to Gulf Coast and Florida. Winters from Gulf Coast to South America.

Whip-poor-will
Caprimulgus vociferus

Robin-sized
Forest habitats

Field Marks
9¾″. A common nightjar of open woodlands and forest, named for its distinctive call. Short, rounded wings without wing bands distinguish it from nighthawks. Nocturnal; usually flies close to ground, feeding on flying insects. Spring through summer, calls continuously at dawn and dusk.
Adult male (1) mottled gray-brown above, buffier below; shows black chin; narrow white necklace; conspicuous white outer tail feathers. Female similar to male but with buff-tipped outer tail feathers; buff necklace. Immatures similar to adults.
A loud, rapid *whip-poor-will*, often repeated for an hour or more.

Similar Species
Nighthawks have long, pointed wings, white wing patches. Chuck-will's-widow larger; call different. Common Poorwill smaller; shorter tail; call different. Common Pauraque has white wing band.

Range
Breeds from south-central Canada east to Maritime Provinces, south to ne. Texas, n. Louisiana, and n. Georgia; also Arizona, New Mexico, and sw. Texas south to Mexico. Winters along Atlantic and Gulf coasts from North Carolina south and in Central America.

Common Pauraque
Nyctidromus albicollis

Robin-sized
Southern Texas: Streamside
Forest

Field Marks
11″. A large gray-and-brown nightjar of forests near watercourses in southern Texas. Like other nightjars, takes flying insects on the wing at night. In flight, white bars on flight feathers are the best field mark, easily visible in car headlights.

Adults (1, 2), have fine pattern of mottled gray and brown; crown grayish, without much mottling; ear patch chestnut; white patch on throat; black spots on shoulders; white bar on flight feathers and white feathers near outer edge visible in flight. Immatures resemble adults.

Call a low, whistled *bur-wheer* or *go-wheer*, slurring downward.

Similar Species
Common Poorwill, *Phalaenoptilus nuttallii* (2), of rocky country in Texas, is smaller, 7½″, with shorter, rounded wings and short tail with white corners. Nightjars have more pointed wings, lack white on tail. Whip-poor-will smaller, grayer; it and Chuck-will's-widow lack white bars on flight feathers. Chuck-will's-widow best told from Common Poorwill by larger size and more extensive white in tail corners.

Range
Resident in s. Texas; also in tropical Central and South America.

Common Nighthawk
Chordeiles minor

Robin-sized
Open Country; Forest habitats;
Residential Areas; Urban Areas

Field Marks
9½″. A widespread and common nightjar found primarily in open and
semi-open habitats, often in cities and towns. Active day and night;
usually seen in evening when swooping overhead to catch insects.
Wings long, pointed, with conspicuous white band; tail long; flight
swift and fluttery. Often in large silent flocks on migration. Nests on
ground or on flat rooftops in cities.
Adult male (1) dark with gray, brown, and white mottling above;
paler below with dark barring; throat white; white band across
primaries conspicuous in flight; tail with white subterminal band.
Adult female similar to male but browner; throat buff; tail without
white band. Immatures similar to female, but throat may be mottled.
A short, nasal *peent* or *pee-eent*, given in flight; booming sound made
by wings in male's territorial display flight.

Similar Species
Lesser Nighthawk has wing band closer to tip; buff in female. Other
nightjars in range have rounded wings. Antillean Nighthawk,
Chordeiles gundlachii, of southern Florida, best told by call, a dry *killy-
ka-dick*.

Range
Breeds across s. Canada south throughout United States. Winters in
South America.

Chimney Swift
Chaetura pelagica

Sparrow-sized
Residential Areas; Urban Areas;
Open Country

Field Marks
5¼". A common, dark, short-tailed swift of suburbs, cities, and open
countryside. Always feeds on the wing, swooping through the air to
take flying insects, alternating wingbeats with graceful but erratic
glides. Small, weak feet preclude perching on branches or wires. In
migration, large flocks swirl in funnel formation over chimneys,
pouring down through opening to spend night, and clinging to walls
with sharp-nailed toes.
Adults (1, 2) dark sooty brown overall, appearing black in flight;
wings long, pointed, curved back; tail short, spine-tipped. Immatures
resemble adults.
Call a rapid, chattering series of chips.

Similar Species
Only swift usually seen in the East. Swallows and Purple Martin have
broader wings, longer tails, more leisurely flight; often perch on
wires, branches, or poles.

Range
Breeds from North Dakota east to Maine and south to Gulf Coast and
Florida. Winters in South America.

261

Purple Martin
Progne subis

Robin-sized
Open Country; Residential Areas

Field Marks
8″. Our largest swallow; distinctive in flight, with a notched tail and nearly triangular wings. It flaps and sails much more than other swallows, sometimes circling in almost soaring flight. Fond of farms, marshes, forest edges, and other open areas, especially near water. Forms postbreeding roosts, sometimes of thousands, particularly in the East, in July and August. Arrives early in spring, departs in late summer. Nests colonially in East in multistory martin houses; in West, mostly in wooded areas with tree cavities.
Adult males (1) are purplish-black, darker on wings and tail. Females (2) and immatures have light bellies, otherwise grayish underparts, with some purplish iridescence on back. Purplish areas often seem black in distance or poor light.
Song complex, rich, liquid; starting with a few clear notes, ending with twitterings and gurglings. Call note a *tyu* or *swee-swuh*.

Similar Species
Smaller Tree Swallow has bluish or greenish back but clean white underparts. European Starling much stubbier, with shorter, even more triangular wings.

Range
Breeds from British Columbia across s. Canada to Nova Scotia and south to Texas and Florida, sparingly in the Rockies and Alaska. Winters in South America.

Tree Swallow
Tachycineta bicolor

Sparrow-sized
Open Country; Lakes, Reservoirs,
Ponds, and Rivers; Residential
Areas

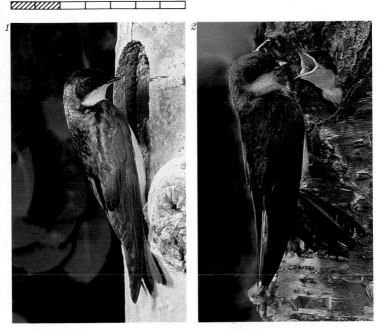

Field Marks
5¾". A gregarious bird that nests in tree cavities near water in
northern areas but also takes to bird houses. Forms large flocks in fall
and winter. Fond of wax myrtle berries in coastal areas. One of the
earliest spring migrants, returning from southern states and Central
America in late winter.
Males (1) have clean white underparts, triangular wings, short,
notched tail; in good light, show metallic light blue or blue-green
upperparts. Females (2) duller above; immatures duller still, almost
brown.
Song a light, high-pitched twittering. Single note a *tsweet*; often given
in flight, sometimes repeated agitatedly.

Similar Species
Brownish-backed immatures similar to smaller Bank Swallow, but
latter has distinct brown breast band. Northern Rough-winged
Swallow has a dusky throat; lacks extension of white from flanks to
rump typical of Tree Swallows.

Range
Breeds from Alaska to Labrador and south to s. California and
Maryland, occasionally to Gulf Coast. Winters from central California
south, along Gulf Coast, and sparingly along Atlantic Coast; also in
Central America.

263

Cliff Swallow
Hirundo pyrrhonota

Sparrow-sized
Open Country; Lakes

Field Marks
5½″. A widespread but local swallow of open country. Less elongated than most swallows, with a squarish tail. Favors bridges, dams, cliffs, farms, and outbuildings, where it plasters its unique, gourdlike mud nests in small colonies. Glides, soars, and circles more than other swallows.
Adults (1, 2) have conspicuous buff rump; forehead ranges from light buff to chestnut; throat chestnut; underparts pale; upperparts dark brown with some bluish-black on back.
Call note a somewhat grating *chwew* or *churr*. Song a series of harsh notes or squeaks.

Similar Species
Cave Swallow, *Hirundo fulva*, of southern Arizona, New Mexico, and Texas, very similar but has a chestnut forehead, darker rump, and light buff throat contrasting with dark cap. Young Barn Swallows with short tails look like Cliff Swallow, but lack the buff rump and light forehead.

Range
Breeds from central Alaska aross Canada to Nova Scotia and south throughout central and ne. United States to Mexico and Texas; absent from Gulf region and southeastern states. Winters in South America.

Barn Swallow
Hirundo rustica

Sparrow-sized
Open Country; Freshwater
Marshes; Salt Marshes

Field Marks
6¾". The familiar Barn Swallow is found nearly worldwide. It is
common, social, and vocal, favoring open country, especially marshes
and farming areas, where it can build its mud nest in buildings
offering protection overhead. A marvelous, extremely agile flyer,
often hawking bugs within inches of the ground; follows mowing
machinery to catch disturbed insects.
Adults of both sexes (1, 2) have dark bluish back, dark wings, long,
deeply forked tail; males have rusty or buff underparts and chestnut
throat and forehead; females and immatures are similar but paler.
Recently fledged young have much shorter tails.
Song complex, combining clear notes, twitters, and gurgles. Call
notes a soft *wit wit* and an emphatic *slip slip*.

Similar Species
Cliff Swallow has similar color patterns but a square-tipped tail and
pale buff rump.

Range
Breeds from Alaska across Canada to Newfoundland south through
United States, but absent inland in Gulf and southern Atlantic Coast
states. Winters in South America.

Northern Rough-winged Swallow
Stelgidopteryx serripennis

Sparrow-sized
Rivers; Streams and Brooks; Open
Country; Salt Marshes;
Freshwater Marshes

Field Marks
5¾″. A light brown, often solitary species, widespread but only
moderately common. Fond of streams, where it nests in burrows in
banks, culverts in bridges, pipes, and similar places, usually in
isolated pairs rather than in colonies. Its flight is slower and more
fluid than that of the smaller Bank Swallow. Returns early in spring,
departs in early fall.
Adults (1, 2) have uniform light brown upperparts and white
underparts; throat has a dusky wash; have short, notched tail visible
in flight.
Call note a rough, slightly grating *brrrtt*, sometimes repeated; harsher
than Bank Swallow's call.

Similar Species
Bank Swallow much smaller, more compact, with a distinct brown
breast band and white throat. Immature Tree Swallow not as brown,
with darker, grayer back; usually has a whiter throat and angular
white areas edging onto rump from upper flanks.

Range
Breeds from se. Alaska across s. Canada to n. New England, south
throughout United States. Winters along Gulf Coast and rarely in
Southwest; also in Mexico.

Bank Swallow
Riparia riparia

Sparrow-sized
Rivers; Streams and Brooks; Open
Country

Field Marks
5¼". Our smallest swallow; a compact brownish bird that flies with
swift, darting flight. A widespread species, locally common near
riverbanks, streams and brooks, road cuts, and other areas with
suitable vertical banks for nesting burrows.
Adults (1, 2) brown above, white below, with a distinct, clean-cut
brown breast band and white throat; the gray-brown rump, a subtle
field mark, contrasts slightly with otherwise uniform brownish
upperparts.
Call a dry, weak chatter or rattle, often noticed before the birds are
seen overhead in migration: *zssaa, zssaa, zssa, zssa*. Other similar
notes given just once or twice.

Similar Species
Northern Rough-winged Swallow larger, lacks breast band, has gray-
brown diffuse wash on throat. Tree Swallow lacks breast band.

Range
Breeds throughout much of North America; absent from
southernmost states. Winters in South America.

Yellow-billed Cuckoo
Coccyzus americanus

*Robin-sized
Eastern Deciduous Forest;
Thickets; Southeastern Pine
Forest; Open Country; Residential
Areas*

Field Marks
12″. An elongated, loose-jointed species, longer but slimmer than a robin. Widespread and fairly common in mixed or deciduous forests, thickets, and streamside woodlands. Fond of caterpillars. Arrives late in spring. Flight swift and direct; often seen flying across highways or between patches of woods. Has more southern distribution than Black-billed Cuckoo, and unlike the latter is also found in the Far West. Rather quiet and slow; perches near tree trunks.
Adults (1) white below, gray-brown above, with a rufous flash in primaries in flight; long tail is dark underneath with large white spots on feather tips; bill mostly yellow, slightly downcurved.
Call a long, staccato *ki-ki-ki-ki-ka-ka-ka-ka-ka-kow-kow-kowlp-kowlp-kowlp*, slowing at end; also a husky, measured *towp, towp, towp, towp*. Sometimes calls on spring nights in migration.

Similar Species
Black-billed Cuckoo lacks rufous primaries, has lighter tail with smaller white tail spots, dark bill, and red eye-ring, visible at close range. Mangrove Cuckoo, *Coccyzus minor* (2), of Florida mangroves and hammocks, similar, but has blackish ear patch, buff breast and belly; shows no rufous in wings in flight.

Range
Breeds from sw. British Columbia east to Maine, south to Mexico and Florida. Local in California. Winters in South America.

Black-billed Cuckoo
Coccyzus erythropthalmus

Robin-sized
Eastern Deciduous Forest;
Thickets

1

Field Marks
12″. A solitary bird preferring more northern and upland forested
areas than the similar Yellow-billed Cuckoo. Slender, longer than a
robin, with a longer tail. Skulks in forests and forages for insects,
chiefly caterpillars.
Adults (1) white below, gray-brown above, with a dark bill, red eye-
ring (visible only at close range), a long, loose tail with small white
tips to tail feathers on underside; sometimes a trace of rufous in the
primaries, but never as richly reddish as in Yellow-billed Cuckoo.
Immatures lack the red eye-ring, are paler under the tail.
Gives a rather rapid *cu-cu-cu* or *cu-cu-cu-cu*, often repeated several
times.

Similar Species
Yellow-billed Cuckoo has a yellow bill, lacks red eye-ring; tail darker
underneath with larger tail spots; primaries rufous. Mangrove Cuckoo
buff below, has black mask, darker underside of tail; chiefly tropical.

Range
Breeds from s. Saskatchewan across s. Canada, south to north-central
Texas, n. Alabama, and Georgia, and along Atlantic Coast to
Maryland. Winters in South America.

269

Belted Kingfisher
Ceryle alcyon

Pigeon-sized
All freshwater habitats; Salt
Marshes; Inshore Waters

Field Marks
13″. A large-headed bird with bushy crest and spear-shaped bill;
frequents streams, rivers, tidal inlets, shorelines, and lakes
throughout North America south of the Arctic. Hovers over small
fish, then plunge-dives headfirst into water. Often perches on exposed
snags and pilings. Excavates nest burrow in steep banks. Flies with
rapid, deep, irregular wingbeats.
Males (1) gray-blue above, white below, with broad gray-blue breast
band. Female (2) has a second rust-colored band on upper belly.
Call a loud, harsh, staccato rattle.

Similar Species
Green Kingfisher, of Arizona and southern Texas, is much smaller,
with green upperparts. Ringed Kingfisher, *Ceryle torquata*, of Rio
Grande in Texas, is larger (16″), with extensive rufous on lower breast
and belly.

Range
Breeds from Alaska east to Newfoundland and south to central
California and Gulf Coast. Winters on open water from se. Alaska east
to Nova Scotia and south to South America.

Green Kingfisher
Chloroceryle americana

Robin-sized
Southern Texas: Rivers; Streams
and Brooks

Field Marks
8¾". A dark green-and-white, long-billed bird of rivers, streams, and brooks in southern Texas; smallest kingfisher in North America. Often perches motionless on branches overhanging a watercourse, waiting to spot small fishes or other prey, then plunges swiftly into water. Common but fairly inconspicuous.
Adult males (1) dark green above with white collar and throat; breast rufous; belly and lower breast white with some dark spotting; in flight, show white spotting on wings and tail. In females, rufous on breast replaced by heavy, dark green spots, which form a band. Immatures resemble females. In all plumages, note heavy, long bill and inconspicuous crest.
Call a sharp *click* or *tick*.

Similar Species
Other kingfishers larger, blue-gray, with more obvious crests; give guttural, rattling calls.

Range
Resident from s. Texas to South America.

271

Budgerigar
Melopsittacus undulatus

Sparrow-sized
Florida: Residential Areas and
Parks; Urban Areas

1

Field Marks
7″. Introduced. Budgies seen in parks and urban areas are escaped
captive birds or their descendants, which have established breeding
populations in wild. Colorful and quite variable, but always easily
recognized as a "parakeet."
Adults (1) mainly bright yellow or yellow-green above, with neat dark
barring; face and throat yellow, with some dark spots under chin;
underparts green; tail long, dark green or blue, and pointed; bill
small, unmistakably parrotlike, curving down to form sharp point.
Immatures less brightly colored, without spots on throat; may have
dark bars on forehead.
Calls include warbles, chatters, and screeches.

Similar Species
Canary-winged Parakeet, *Brotogeris versicolorus*, occurs in parts of
Florida; larger (9″), solid green with bold yellow patches in wings.

Range
Established in Florida along coasts and at scattered sites inland.

Buff-bellied Hummingbird
Amazilia yucatanensis

Very Small
Southern Texas: Streamside
Forest

Field Marks
4¼". A tiny inhabitant of streamside forests and woodland borders in
southern Texas. Hovers before flowers to take nectar, exposing
iridescent green gorget on throat. One of three closely related
Mexican hummingbirds to reach our range; male and female appear
alike.
Adults (1) have emerald-green head and upper breast; wings and back
slightly darker green; underparts pale buff; tail rufous; bill long, red
with black tip.
Gives high, squeaky, metallic notes.

Similar Species
No other eastern hummingbird has entirely rufous tail.

Range
Resident but scarce in s. Texas; occasionally wanders to northern
Texas coast and s. Louisiana in winter. Chiefly in Mexico and Central
America.

Rufous Hummingbird
Selasphorus rufus

Very Small
*Gulf Coast: Residential Areas and
Parks*

Field Marks
3¾". Chiefly a western species, although small numbers appear each
winter in Texas and Louisiana. Found in gardens, meadows,
woodlands, and parks. Visits hummingbird feeders.
Males (1) bright rufous above, on sides, and top of tail; breast white;
throat (gorget) bright iridescent orange-red; crown green. Occasional
males have some green on back. Females (2) and immatures are
whitish below; rufous on sides, rump, and most of tail; greenish on
upper wings and back.
Calls a *tchup* and an excited, buzzy *zee-tchuppity-tchup*.

Similar Species
Female Ruby-throated Hummingbird (usually absent from United
States in winter) lacks rufous in tail and on sides.

Range
Breeds from s. Alaska to s. Oregon and w. Montana. Winters in
Mexico and occasionally in s. California. Wanders annually to Gulf
Coast, from Texas east to Florida.

Ruby-throated Hummingbird
Archilochus colubris

Very Small
Forest habitats; Residential Areas
and Parks; Open Country

Field Marks
3¾". The only hummingbird likely to be seen in most of its range.
Often first detected by the dull, droning hum made by buzzing wings.
Somewhat common in gardens, woods, parks, and along hedgerows;
often seen taking off across broad expanses of water, or feeding
among trumpet-creeper flowers (bignonias).
Males (1) green above, mostly white below, with a red throat (gorget)
that appears dark or black in poor light; tail dark. Females (2) and
immatures lack the red gorget; mostly greenish above, white on
throat and underparts; have rounded tails with white spots; some
immatures have dark marking or a trace of the red gorget.
Call note a soft, subtle *tchupp*, sometimes repeated excitedly two to
five times, becoming squeakier and higher-pitched.

Similar Species
Male Black-chinned Hummingbird, *Archilochus alexandri*, of western
edge of Great Plains, has black gorget with violet throat; immatures
and females of these two species indistinguishable in the field. Rufous
Hummingbird much more reddish.

Range
Breeds from Alberta east to Nova Scotia and south to e. Texas and
Florida. Winters from Mexico to Costa Rica, rarely along Gulf Coast.

Verdin
Auriparus flaviceps

Very Small
Texas: Streamside Forest; Dry
Western Woodlands; Open
Country

Field Marks
4½". A small, active, titlike bird of scrubby areas, streamside forest, mesquite, palo verde, and dry brush. Makes bushy nest, also used as a roost. Flits around in bushes, making quick, erratic flights.
Adults (1, 2) have upperparts and wings uniform gray-brown with chestnut patch on the shoulders; uniform light gray below; head and throat yellow. Juveniles are browner; lack chestnut shoulder and yellow head and throat.
Song a loud *tsee tsee tsee*, first note higher. Call an emphatic *tszee* or *tszip*.

Similar Species
Northern Beardless-Tyrannulet, *Camptostoma imberbe*, a warblerlike flycatcher resident in thickets and streamside forest along Mexican border in extreme southern Texas, Arizona, and New Mexico, has wing bars, suggestion of eye-ring, and pale lower mandible.

Range
Se. California east to s. Texas and south to Mexico.

Boreal Chickadee
Parus hudsonicus

Sparrow-sized
Boreal Forest

Field Marks
5½". A chickadee indelibly associated with the northern coniferous
forest. A few in some winters stray south from their normal sub-arctic
haunts. Widespread in boreal spruce forests.
Adults (1, 2) are brownish or dusky brown, with a dark cap, rufous
sides and flanks, black bib, dirty white central belly and breast;
brown-backed, with a grayish area around the ears, and small whitish
cheek patches.
Utters a wheezy, husky *chichee-day-day*, completely unlike clear
whistled call of Black-capped Chickadee.

Similar Species
Black-capped Chickadee much paler, has black cap; clear, whistled
calls; large white cheeks.

Range
Alaska east across Canada to Newfoundland, and south to n. United
States.

Black-capped Chickadee
Parus atricapillus

Sparrow-sized
Forest habitats; Residential Areas
and Parks

Field Marks
5¼″. A small, very active, and common bird; a frequent visitor to bird feeders. Tame and inquisitive; its arrival in woods, thickets, brush, and residential areas usually signals the presence of other forest birds that accompany chickadees in mixed-species foraging flocks.
Adults (1) have black bib and cap, large white cheeks, plain gray upperparts, whitish underparts; tinge of rust on flanks and sides. Immatures similar.
Call a dry *chick-a-dee-dee* or *dee-dee-dee*. Song a simple, clear, whistled *dee dee*, first note higher.

Similar Species
Carolina Chickadee, *Parus carolinensis* (2), found in same habitats from Kansas and New Jersey south to Gulf Coast and Florida, lacks well-defined pale edges on middle of folded wing; has higher-pitched calls, and four-noted song: *fee-bee, fee-bay*. See Boreal Chickadee.

Range
Alaska east to Newfoundland and south to n. California, n. Oklahoma, and n. New Jersey.

Tufted Titmouse
Parus bicolor

Sparrow-sized
Forest habitats; Residential Areas
and Parks

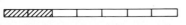

Field Marks
6½″. A common inhabitant of eastern forests. Outside breeding
season, often seen with other small birds, such as chickadees, in
residential areas and parks. Conspicuous despite small size and gray
color; forages actively at every forest level. Very vocal; often in small
flocks. A black-crested form occurs in Texas.
Adults (1) gray above, with conspicuous gray crest and some black on
forehead; underparts paler gray; flanks buff. Immatures resemble
adults. "Black-crested" race (central and southern Texas) has black
crest, gray forehead.
Song consists of two loud *peter* notes repeated many times; call a
harsh *see-jert*.

Similar Species
Dark-eyed Junco larger, much rounder, with distinctive pink bill;
lacks crest; usually forages on ground.

Range
Resident from Nebraska east to s. New England, south to n. Mexico
and Florida.

Olive-sided Flycatcher
Contopus borealis

Sparrow-sized
Boreal Forest; Eastern Deciduous
Forest

Field Marks
7½". A fairly large, rather short-tailed flycatcher; large-headed, with heavy bill. Most often seen perched on the tip of dead snags or at the very top of a tree. Inhabits northern coniferous forests, mountainous forests in the West, and deciduous forests of the East.
Adults (1, 2) drab olive-brown above and on sides; white or off-white throat, central breast, and belly create "vested" appearance below; head large; bill long, stout, dark. Immatures similar to adults but generally darker above and below.
Song an emphatic, clear whistle: *quick-three-beers*, with second note highest; also a *pip-pip-pip* call.

Similar Species
Wood-pewees paler and slightly smaller, with wing bars, and paler underparts without vested look.

Range
Breeds from Alaska east to Newfoundland and south to s. California and w. Texas, Great Lakes region, and n. New England. Winters in South America.

Eastern Wood-Pewee
Contopus virens

Sparrow-sized
Forest habitats; Groves

Field Marks
6¼". A small gray bird of forest habitats; most commonly seen
flycatching for insects, darting out from a perch to snatch a flying
insect from the air, then doing a quick reverse back to the perch.
Found throughout the East in groves and woodlands, where it forages
high in trees. The nearly identical Western Wood-Pewee can be
distinguished only by voice.
Adults (1, 2) grayish-olive or olive-brown above, with two white wing
bars; bill thin and roundly pointed; tail dark, fairly long; underparts
whitish to pale gray. Immatures similar, with buff wing bars.
Song a long series of whistled phrases, *pee-wee*, *pee-a-wee*, repeated
many times.

Similar Species
Western Wood-Pewee, *Contopus sordidulus*, of western edge of
Great Plains, indistinguishable except by voice: a burry *peeyeer* or
peeyeet. Eastern Phoebe lacks wing bars; frequently bobs tail.
Empidonax flycatchers smaller; usually have bold eye-ring and more
conspicuous wing bars. Olive-sided Flycatcher larger, darker, with
dark "vest."

Range
Breeds from south-central Canada east to Nova Scotia and south to
Gulf Coast and Florida. Winters in Central and South America.

Eastern Phoebe
Sayornis phoebe

Sparrow-sized
Streams and Brooks; Forest
habitats

Field Marks
7″. A dark, medium-sized flycatcher, commonly seen in the East, and
easily recognized by its upright posture and habit of bobbing its tail
when perched. Found in forested areas, usually near streams, brooks,
and small rivers.
Adults (1, 2) dark grayish above with pale gray or whitish underparts;
head dark, with no eye-ring; tail long, dark; wings uniformly dark.
Immatures similar to adults, but have prominent buff bars on wings
and pale yellowish tone to underparts.
Song a clear *fee-bee, fee-bay*, repeated many times; also gives a clear
chip note.

Similar Species
Wood-pewees paler and slightly smaller, with wing bars, and without
contrasting dark head or eye-ring.

Range
Breeds from central Canada east to New Brunswick, south to central
Texas and Maryland. Winters from s. Oklahoma east to Delaware and
south to Mexico and Florida.

Least Flycatcher
Empidonax minimus

Sparrow-sized
Eastern Deciduous Forest;
Groves; Open Country;
Residential Areas

Field Marks
5¼″. The most familiar *Empidonax* flycatcher in East, found in
deciduous forests, groves, and in large shade trees in residential
areas, where persistent call is often best clue to its presence. Usually
forages higher in trees than Yellow-bellied, Willow, and other small
flycatchers of this group.
Adults (1, 2) are small, large-headed flycatchers; gray-green above,
paler below, with two pale wing bars and distinct, though narrow,
eye-ring.
Call a dry, repeated *chebec, chebec.*

Similar Species
Willow and Alder flycatchers have faint eye-rings; Alder and Acadian
more greenish above; Yellow-bellied and immature Acadian have
yellow underparts. All these species best told by voice. Wood-pewees
larger, darker, with less obvious wing bars and no eye-rings.

Range
Breeds across s. Canada east to Nova Scotia, south to central
Wyoming and New Jersey, in the mountains to n. Georgia. Winters
from Mexico to South America.

Yellow-bellied Flycatcher
Empidonax flaviventris

Sparrow-sized
Boreal Forest; Eastern Deciduous
Forest

Field Marks
5½". A small, quiet, large-headed flycatcher of northern coniferous forests, where it frequents spruce bogs. Usually found in deciduous forest during migration. Often forages at eye level, but may be shy and difficult to see.
Adults (1) greenish above with two pale wing bars; white eye-ring (not always obvious); yellow on throat and rest of underparts.
Call a sharp *killick* or soft *chuwee?*

Similar Species
Acadian Flycatcher, *Empidonax virescens* (2), summer resident in deciduous forest from Great Lakes region to Gulf Coast and central Florida, is whiter below, but fall immatures can be yellowish on underparts (except for throat); best told by song, a piercing *peet-seet!* or *per-reet!* See Least and Willow flycatchers; best told by call. Wood-pewees larger, darker, with less obvious wing bars and no eye-rings.

Range
Breeds from nw. Canada, Hudson Bay, and Newfoundland south to n. Great Lakes region and n. New England. Often seen in migration in s. United States. Winters in tropics.

Willow Flycatcher
Empidonax traillii

Sparrow-sized
Moist Thickets; Brushy Open
Country; Groves

Field Marks
5¾". A small, inconspicuous flycatcher of willow swamps, moist thickets, woodland edges, and margins of ponds and streams. Often flicks tail upward nervously.
Adults (1) are brownish-olive above, whitish below, with two white wing bars and very weak eye-ring.
Song a wheezy *fitz-bew*, accented on first syllable.

Similar Species
More northern Alder Flycatcher, *Empidonax alnorum* (2), summer resident in alder swamps and aspen groves in Alaska, Canada, Great Lakes region, and New England, is almost identical, but may have a slightly more pronounced eye-ring; best told by song, a burry *wee-be-o*, and nesting habitat. Least, Yellow-bellied, and Acadian flycatchers have bolder eye-rings; Least grayer above, Yellow-bellied and Acadian greener; these species best told by voice and nesting habitat. Wood-pewees larger, darker, with less obvious wing bars and no eye-rings.

Range
Breeds from s. British Columbia, s. Ontario, Great Lakes region, and s. New England south to central California, the Southwest, Arkansas, Ohio, and North Carolina. Winters in tropics.

Ash-throated Flycatcher
Myiarchus cinerascens

Robin-sized
Texas: Streamside Forest; Open
Country

Field Marks

8½". A grayish-brown, crested flycatcher of deserts with mesquite
and cactus, dry western woodlands, and thickets near watercourses in
south Texas. Generally forages in shrubby areas, flying out to take
insects on the wing, but also taking crawling insects.

Adults (1, 2) grayish-brown above, with bushy crest and some white
edgings on wings; pale gray on throat and breast; belly has pale
yellow wash; underside of tail brown with rufous wash. Immatures
similar with more red in tail. In all plumages, bill is fairly thin
and small.

Song a series of *ha-wheer* notes. Call a *ka-brick* or single *ha-wheer*.

Similar Species

Great Crested and Brown-crested flycatchers best distinguished by
voice; generally darker above with more extensive yellow below.
Brown-crested somewhat larger.

Range

Breeds from Washington east to Colorado and south to central
Mexico. Winters from Arizona south to Costa Rica, occasionally along
Gulf Coast. May wander farther east.

Brown-crested Flycatcher
Myiarchus tyrannulus

Robin-sized
Southern Texas: Streamside
Forest

Field Marks
8¾". A fairly common and conspicuous inhabitant of dry woodlands
and desert areas with cactus in the Southwest and southern Texas.
Large and fairly drab, but often easily detected by its loud calls, and,
in spring, by its distinctive song. Formerly known as Wied's Crested
Flycatcher.
Adults (1) gray-brown above, with bushy crest on head; throat and
breast pale gray; belly whitish; tail and wings have rufous tones,
wings have white wing bars; bill large, heavy, and black. Immatures
show more red in tail.
Song a whistled, musical series of *whit* notes alternating with *burr-rr*,
whee-rr, and *purreeeer* notes; call a sharp *whit!*

Similar Species
Great Crested Flycatcher almost identical; safely distinguished only
by loud *wheeep!* call and song. At very close range, pale lower
mandible, slightly more olive upperparts, somewhat darker throat and
breast, and brighter belly may be noticeable. See Ash-throated
Flycatcher, also best told by voice.

Range
Breeds from s. California east to s. Texas. Winters in Mexico and
Central America, occasionally in s. Florida.

Western Kingbird
Tyrannus verticalis

Robin-sized
Open Country and Grasslands;
Groves

Field Marks
8¾″. A solidly built, elongated kingbird with a heavy bill. Abundant in open areas and grasslands with sufficient scattered trees to offer perching posts; also partial to farmland with fences and posts. Hawks for insects, as do other flycatchers. Although chiefly a bird of the Great Plains and Far West, it occurs regularly in the East during fall and winter.
Adults (1, 2) have gray head, wings, and back, with greenish wash on back; dusky white throat and gray upper breast; pale lemon-yellow lower breast and belly; tail black with white outer feathers.
Call a sharp, far-carrying *kit* or *whit*; also gives a flight song and other chattering notes.

Similar Species
Eastern Kingbird has white underparts. Couch's Kingbird lacks white on outer tail feathers; has notched tail.

Range
Breeds from s. British Columbia east to Wisconsin and south to California and Texas. Winters in Central America, occasionally in Florida. Small numbers on Atlantic Coast in fall.

Great Crested Flycatcher
Myiarchus crinitus

Robin-sized
Eastern Deciduous Forest;
Residential Areas; Open Country

Field Marks
8¾". A large, aggressive, vocal flycatcher, more often heard than
seen. Feeds high in the canopy or in tops of trees in wooded
residential areas; also occurs in suitable areas in open countryside.
Despite name, most birds do not have an obvious crest.
Adults (1, 2) brownish above, with pale yellow lower breast and belly;
throat and upper breast gray; outer flight feathers and inner tail
feathers marked with rich cinnamon-brown.
Song a loud, whistled *wheeerp* or *whee-err*; also gives a loud, harsh
wheep call.

Similar Species
Brown-crested Flycatcher paler gray-brown above, with somewhat
paler throat and breast; belly slightly duller yellow. Ash-throated
Flycatcher paler gray (almost whitish) on throat and breast, with
paler yellow belly. All three species best distinguished on basis of
songs and calls.

Range
Breeds from south-central Canada east to New Brunswick and south
to Texas and Florida. Winters in s. Florida, Mexico, and South
America.

Couch's Kingbird
Tyrannus couchii

Robin-sized
Southern Texas: Streamside
Forest

Field Marks
9¼". A noisy and aggressive flycatcher, fairly common in summer in streamside forests and thickets in southern Texas; prefers chaparral with mesquite and persimmon. Usually sings at dawn. Until recently, was considered the same species as Tropical Kingbird.
Adults (1) have greenish back, brownish wings and tail, and yellow underparts; bill large; tail notched; throat very pale gray to white; ears have darker gray patch. Immatures similar to adults.
Song a buzzy *queer*, *chi-queer*; call a short *kip*.

Similar Species
Western Kingbird has fan-shaped black tail with white outer edges. Ash-throated, Great Crested, and Brown-crested flycatchers much browner above, with long, straight, unforked tails; Ash-throated and Brown-crested may have much less yellow on underparts.

Range
Resident in s. Texas; also in Central America.

Great Kiskadee
Pitangus sulphuratus

Robin-sized
Southern Texas: Streamside
Forest

Field Marks
9¾". A heavy-billed, stocky flycatcher with bold black-and-white head
pattern; enters our range only in southern Texas. Noisy and active,
especially at dusk and dawn; fairly common in streamside forested
areas, thickets, orchards, groves, and residential areas. Known to
dive into streams to take small fishes.
Adults (1) have white eyebrows and throat, with broad black mask
crossing from base of bill through eyes to nape; back olive-brown;
wings and tail rufous; underparts yellow. Immatures resemble adults.
Song a loud *kis-ka-dee*.

Similar Species
No other flycatcher in our range has Great Kiskadee's bold black-and-
white head pattern.

Range
Resident from s. Texas south to Argentina. Occasionally occurs in
other southern states.

Gray Kingbird
Tyrannus dominicensis

Robin-sized
Mangroves; Forest habitats;
Residential Areas

Field Marks
9″. A large gray-and-white flycatcher of Florida and the Gulf Coast;
found only in wooded coastal areas, such as mangrove forests,
swamps, and nearby wooded residential areas. Often perches on
telephone wires and exposed branches to watch for insects; will also
take fruits and berries of some tropical trees.

Adults (1, 2) have gray head, back, wings, and tail; underparts very
pale gray to whitish; bill long, heavy-looking; tail black, with shallow
notch at tip. Immatures similar to adults.

Song a loud *pit-cheer-ree* or *pit-cheer-ry*.

Similar Species
Eastern Kingbird blackish above, much whiter below, with white on
tip of black tail. Other flycatchers in same range either have yellow
underparts or much browner upperparts; most do not have notched
tip of tail.

Range
Breeds along Atlantic and Gulf coasts from South Carolina to
Louisiana. Winters in West Indies and South America.

Eastern Kingbird
Tyrannus tyrannus

Robin-sized
Open Country; Residential Areas;
Forest habitats

Field Marks
8½". A large, distinctive, handsome flycatcher, found in open country, agricultural areas, woodland edges, and along streams and brooks. Vocal and active, sallying forth from a perch on a tree, telephone post, or overhead wire to catch insects on the wing.

Adults (1, 2) unmistakable; head and face coal-black; throat and underparts white, forming a partial collar at neck; upperparts gray with vague white feather edgings on wing and two indistinct wing bars; tail blackish with conspicuous white tip.

A strident, harsh series of *killy-killy-killy* notes, often ascending; also gives *dzee, dzee* notes and squeaky chatters.

Similar Species
Gray Kingbird, of southern and coastal Florida, has notched black tail, grayer upperparts; different call. Western Kingbird has yellow wash on lower belly, paler gray head and upperparts; white feathers on outside of tail.

Range
Breeds from British Columbia east to Nova Scotia and south to ne. California, the Gulf Coast, and Florida. Winters in South America.

293

Scissor-tailed Flycatcher
Tyrannus forficatus

Robin-sized
Open Country and Grasslands

Field Marks
13″. A fairly tame, attractive bird of open areas, farmland, and plains with scattered trees and fence posts. Very much like a kingbird in appearance, but with an extremely long and elegant tail. In spring, males perform elaborate "sky-dance" courtship ritual, spiraling to great heights and then plunging downward, then zigzagging up and down, calling all the while.

Adults (1) pale gray above, palest on the head, with black wings; underparts gray with pale pinkish suffusion on sides; tail black; outer tail feathers extremely long, patterned in black and white. Females usually somewhat smaller and shorter-tailed than males. Immatures dull brownish above, pale below; short black tail somewhat notched at tip, with white outer edges; sides washed with pink.

Call a harsh *kip* or *kit*; also chattering and twittering notes; in general, vocalizations like those of Western Kingbird.

Similar Species
Western Kingbird similar to immature Scissor-tailed, but has lemon-yellow on belly and sides.

Range
Breeds from s. Nebraska south to Texas and w. Louisiana. Winters in Mexico and Central America, and regularly in s. Florida.

Say's Phoebe
Sayornis saya

Sparrow-sized
Southern Texas: Open Country

Field Marks
7½". A brownish flycatcher of dry, open country and areas with
scattered vegetation. Perches on dead twigs and branches as well as
fence posts; often pumps its tail up and down when perched. Highly
migratory, frequently appearing briefly in places far outside its usual
range and habitat.
Adults (1, 2) brownish-gray above and on upper breast; lower breast
and belly cinnamon-rufous; throat pale gray to whitish; bill long,
heavy, black; tail dark; wings have indistinct narrow bars. Immatures
browner above, with two rufous wing bars.
Song a rapid *pit-tsee-ar*; call a mellow whistle: *pee-urrr*.

Similar Species
Female Vermilion Flycatcher paler gray above, nearly whitish on
breast, with dusky streaks; more wine-colored wash to lower belly.
Wood-pewees lack rufous on belly. Eastern Phoebe has darker head,
whitish throat, rest of underparts whitish (or yellowish in immatures),
with no buff or cinnamon.

Range
Breeds from e. Alaska east to Saskatchewan and south to Mexico.
Winters from central California, s. Utah, and s. Texas to Mexico,
locally from farther north.

Rose-throated Becard
Pachyramphus aglaiae

Sparrow-sized
Southern Texas: Streamside
Forest

Field Marks
7¼". An unmistakable but rarely encountered species with a large
head and a stocky build. North American range limited to areas of
streamside vegetation in parts of Arizona and southern Texas.
Usually perches near tree trunks; forages in dense cover, moving
slowly about in search of insects. Best clue to its presence usually its
chattering call.
Adult males (1) have dark head, gray upperparts, pale gray
underparts, and rose-colored patch on throat. Adult females (2) have
similar pattern, but grays replaced with dark brown above and buff
below; lack throat patch; have buff collar and rufous-buff cheek patch;
cap black. Immatures similar to females; young males may be more
gray than brown and buff, with hint of throat patch.
Call a loud series of high, rapid chattering notes, followed by a high
whistle: *seee-oo.*

Similar Species
Immature Scissor-tailed Flycatcher similar to females and immatures,
but has pinkish wash on sides; black tail has white outer feathers.

Range
Breeds very locally along streams in se. Arizona and Rio Grande
valley of Texas. Winters mainly in Central America.

Vermilion Flycatcher
Pyrocephalus rubinus

Sparrow-sized
Texas: Streamside Forest; Open
Country

Field Marks
6″. Unmistakable. A small, brightly colored flycatcher usually found in woods or brush along streams or near ponds and stock tanks in dry, open country. May hawk for flying insects from a perch only a few inches from ground. Often bobs tail like a phoebe.
Adult males (1) have whole crown and underparts bright scarlet, with line through eye and entire upperparts dark brown. Females (2) brownish above, with pale eyebrow, dark mask through eye and in ear region, whitish underparts faintly streaked with brown, and buff or pinkish undertail coverts. Young male resembles female, but has red on belly.
Call a sharp *pit-zee* or *pits*.

Similar Species
Immature Say's Phoebe may show yellowish or buff on belly, but lacks whitish throat and breast of female Vermilion Flycatcher.

Range
Breeds from se. California east to central Texas and south to Argentina. Winters over much of breeding range; wanders widely.

Red-headed Woodpecker
Melanerpes erythrocephalus

Robin-sized
Forest habitats; Residential Areas
and Parks

Field Marks
9¼″. A boldly patterned, sociable woodpecker of open woodlands and areas with scattered trees. Especially common where there are oaks and hickories; stores nuts in crevices in bark. Often seen perched conspicuously on tips of dead branches.

Adults (2) have whole head red; back, tail, and most of wings black; rump, wing patch, and underparts white. Immature (1) has black and red areas of adult replaced with brown, and two dark bars across white wing patch.

Calls a loud, rattling *kweer*, and various chattering notes.

Similar Species
No other medium-sized woodpecker has bold white patches on trailing edge of wings.

Range
Breeds from s. Alberta, s. Ontario, and Quebec south to ne. New Mexico, Texas, Gulf Coast, and Florida. Winters north to Kansas and New Jersey.

Pileated Woodpecker
Dryocopus pileatus

Crow-sized
Forest habitats

Field Marks
16½". A large, crested woodpecker of forested and wooded areas, occasionally visiting residential areas with large trees. Usually quite shy; presence best detected by loud call.

Adults mainly black, with black and white stripes on face, white stripe down side of neck, white bases to primaries, and white wing linings. Male (2) has bright red crest, crown, forehead, and mustache; female (1) has red only on crest and crown.

Call a loud, rolling *kuk-kuk-kuk-kuk-kuk-kuk*, often dropping in pitch.

Similar Species
Ivory-billed Woodpecker, *Campephilus principalis* (probably extinct), inhabits southeastern swamps; larger, with very bold white wing patches, whitish bill.

Range
Resident from n. British Columbia, Ontario, and Quebec south to n. California, Montana, e. Nebraska, e. Texas, Gulf Coast, and Florida.

Yellow-bellied Sapsucker
Sphyrapicus varius

Robin-sized
Forest habitats; Residential Areas
and Parks

1
2

Field Marks
8½". A furtive, medium-sized woodpecker of deciduous and mixed
forests, orchards, and residential areas. Drills rows of small holes in
bark, then feeds on sap and insects attracted to it. Mottled pattern
makes resting bird difficult to see.
Adults have red crown, black and white stripes on face, and barred
back; patch of white in wing and white rump are conspicuous in flight;
underparts pale yellow with black crescent on breast. Male (2) has
red throat, female (1) has white throat. Immature has brownish head
and back.
Call a whining *churrr* or *skeeer*.

Similar Species
Other woodpeckers are more distinctly patterned in black and white,
lack mottled effect of sapsuckers.

Range
Breeds from Alaska, Ontario, and Newfoundland south to ne. British
Columbia, e. North Dakota, Missouri, Pennsylvania, and Connecticut.
Winters from Missouri, Ohio Valley, and New Jersey south to Gulf
Coast and Florida.

Hairy Woodpecker
Picoides villosus

Robin-sized
Forest habitats; Residential Areas
and Parks

Field Marks
9¼″. A medium-sized black-and-white woodpecker of forests and
woodlands; often visits gardens and residential areas, but less
frequently than Downy Woodpecker. Forages mainly on trunks and
large branches of trees. Shier than the Downy.
Adults have black-and-white-striped head, black upperparts with
white in center of back and white spots on wings, and white
underparts; outer tail feathers white, without narrow black bars.
Males (1) have small red patch on nape, lacking in females (2); young
birds may show red on crown.
Calls a loud, high-pitched *peek!* and a short rattle.

Similar Species
Downy Woodpecker similar but smaller, with shorter bill, barred
outer tail feathers, and softer calls. See Three-toed and Black-backed
woodpeckers.

Range
From s. Alaska east to Newfoundland and south to Central America
and Florida. Some northern birds move south in winter.

Downy Woodpecker
Picoides pubescens

Sparrow-sized
Forest habitats; Residential Areas
and Parks

Field Marks
6¾". A small, widespread, and abundant woodpecker with a very
short bill, found in most wooded habitats and in suburban areas.
Readily visits feeders, and often joins mixed winter flocks of
chickadees, nuthatches, and kinglets. Usually forages on smaller
branches than similar Hairy Woodpecker.
Adult has black-and-white-striped head, black upperparts with white
in center of back and white spots on wings, and white underparts;
outer tail feathers white with narrow black bars. Male (2) has small
red patch on nape, lacking in female (1).
Calls a high-pitched *pik* and a rattling *ki-ki-ki-ki-ki*.

Similar Species
Hairy Woodpecker similar but larger, with longer bill, pure white
outer tail feathers, and louder, more piercing calls. See Ladder-
backed Woodpecker.

Range
From se. Alaska east to Newfoundland and south to California and
Florida. Northern birds move south in winter.

Red-cockaded Woodpecker
Picoides borealis

Robin-sized
Southeastern Pine Forest

Field Marks
8½". An endangered, medium-sized woodpecker of mature pine
forests, where it is rare, very local, and found in small colonies.
Trunks of trees with large patches of bark scaled off, or nest holes
surrounded by oozing sap, may be signs of bird's presence.
Adult (1, 2) has black-and-white barred back, white underparts with
streaked flanks, black crown, and conspicuous white cheek patches;
small red patch, or cockade, behind eye of male is seldom visible.
Calls include a flycatcherlike *skrrit* and various rasping notes.

Similar Species
Downy and Hairy woodpeckers have solid white backs, lack bold
white cheek patch, and have different calls; Downy much smaller.

Range
Locally from se. Oklahoma east to Virginia and south to e. Texas and
Florida.

Black-backed Woodpecker
Picoides arcticus

Robin-sized
Boreal Forest

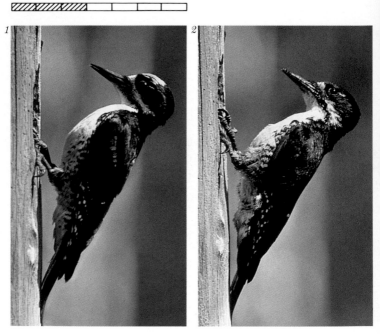

Field Marks
9½". Inhabits coniferous boreal forests, especially spruce bogs. Often seen working low in dead trees. Less noisy and conspicuous than most woodpeckers, often working relatively quietly and flying from tree to tree silently, though probably noisier than Three-toed, with which it occurs in Canada and the extreme Northeast.

Adult male (1) glossy black above with yellow patch on crown, black-and-white mustache; underparts white, barred with black on flanks and sides; wings sparsely barred with black and white; tail black with white outer feathers. Adult female (2) similar but lacks yellow on crown. Immatures duller, often with yellow crown spot like adult male.

Call a sharp *kyik*, similar to Hairy's but flatter.

Similar Species
Hairy and Downy woodpeckers lack barred sides, have white on back. Three-toed Woodpecker has variably barred white back, white eyebrow.

Range
Resident from central Alaska south to central California and across United States and Canada to Newfoundland and Maine.

Three-toed Woodpecker
Picoides tridactylus

Robin-sized
Boreal Forest

Field Marks
8¾". A woodpecker of northern coniferous forests, especially spruce-fir, where it frequents areas with dead trees such as spruce bogs and burns. Generally relatively shy, quiet, and inconspicuous, working low on trees.

Adult male (1) has blackish upperparts with yellow crown and white eyebrow behind eye curving downward; white mustache; and white back variably barred with blackish; underparts white, heavily barred with black on sides and flanks; wings barred with black and white; tail black with white outer tail feathers. Adult female (2) similar, but lacks yellow. Juveniles dull black; most have yellow spot on crown.

Call a sharp *pik*, slightly lower than Downy's. Also a dry rattle.

Similar Species
See Black-backed Woodpecker. Hairy and Downy woodpeckers lack barred sides and flanks.

Range
From s. Alaska east to Newfoundland and south in mountains to New Mexico and Arizona, and n. Minnesota to n. New England.

Golden-fronted Woodpecker
Melanerpes aurifrons

Robin-sized
Texas and Oklahoma: Streamside
Forest; Dry Western Woodlands;
Groves; Residential Areas

Field Marks
9¾". A common, conspicuous, and noisy resident in a variety of
woodland habitats in central and southern Texas and southern
Oklahoma. An attractive woodpecker, often seen on poles and trees
along roadsides.
Adult males (2) tan-gray on head and underparts, except for golden
patches on forehead, nape, and belly, and scarlet crown patch; back
and wings heavily barred black and white, with white rump and white
patches in outer wing; tail black; black-and-white barring on outer
feathers. Adult female (1) similar but lacks scarlet patch. Juvenile
similar but lightly streaked on breast; head brownish.
Call a harsh, rolling *churrr*, given singly or in series. Also a rapid,
nasal *yuk-yuk-yuk*, or *ek ek ek ek*; generally more harsh than Red-
bellied's call.

Similar Species
Red-bellied Woodpecker has black-and-white-barred tail; male has
extensive red on forehead, crown, and nape; female has red patch on
nape; both sexes have pale reddish patch on lower belly. Northern
Flicker larger, browner above, spotted and streaked below.

Range
Resident from central Texas and sw. Oklahoma south to Mexico; also
in Central America.

Red-bellied Woodpecker
Melanerpes carolinus

Robin-sized
Eastern Deciduous Forest;
Southeastern Pine Forest;
Residential Areas and Parks

Field Marks
9¼". A widespread, common woodpecker throughout the East, with black-and-white-patterned upperparts and red nape. Noisy and fairly aggressive; frequents residential areas, parks, and even feeders, as well as pine and deciduous forests.
Adult males (2) have black-and-white-barred back and wings; head red; cheeks and underparts plain whitish, except for faint reddish wash on lower belly. Females (1) similar, but have grayish-white crown and forehead, and tiny area of red feathering at base of bill. Immatures have streaking on breast and dark crown. In flight, all plumages show white patches on wings and white rump.
Usual call a rattling *churr.*

Similar Species
In Texas and Oklahoma, see Golden-fronted Woodpecker. Male Ladder-backed Woodpecker has much less red feathering on head, heavy spotting on underparts. Red-cockaded, Hairy, and Downy woodpeckers have white patches on head and dark facial markings; underparts have spots and streaks; lack distinctive, neat zebra markings of Red-bellied.

Range
Breeds from central South Dakota east to sw. Connecticut and south to central Texas and Florida. Northern birds move south in winter.

Northern Flicker
Colaptes auratus

Pigeon-sized
Forest habitats; Open Country

Field Marks
12½". A common and widespread, rather large woodpecker. Found in
most habitats with trees, including gardens and parks. Noisy and
conspicuous; frequently seen hopping on ground in search of ants.
Flies rather slowly, often above treetops; wingbeats somewhat
irregular, slow, and deliberate; flashes color of wing linings.
Eastern form usually called "Yellow-shafted" Flicker; male (2) has tan
head, gray crown, red nape, and black mustache; back and
upper surface of wings barred grayish-brown and black; rump white;
black crescent on breast; rest of underparts light tan, heavily spotted
with black; top of tail black, underside yellow; underwings uniform
yellow; dark gray bill large, long, and chisel-shaped. Female (1) lacks
mustache. "Red-shafted" Flicker (western form) sometimes occurs in
eastern range; has brown crown and pinkish-orange wing linings.
Calls include a nasal, long, loud series, *wik-wik-wik-wik*; louder,
faster, longer, and higher-pitched than similar call of Pileated
Woodpecker. Also a piercing, sharply descending whistled *peeahr*.

Similar Species
No other woodpecker is barred with gray-brown and black, with
white rump.

Range
Breeds from Alaska east to Quebec and south throughout United
States. Northern birds move south in winter.

Ladder-backed Woodpecker
Picoides scalaris

Sparrow-sized
Texas and Oklahoma: Dry
Western Woodlands; Thickets;
Residential Areas and Parks

Field Marks

7¼″. A little black-and-white woodpecker; noisy and conspicuous as it works dead limbs and trunks of trees and shrubs. Often seen on much smaller limbs than other woodpeckers.
Male (1) has red crown patch, white face; narrow black line runs from eye back to middle of head, where it curves sharply down and then back toward the bill; back and wings barred with black and white; underparts off-white, spotted on sides and flanks; tail black, with white outer tail feathers spotted with black. Female (2) similar but lacks red.
Call a sharp *peek*, similar to Hairy Woodpecker's but higher. Also a descending, rattling whinny.

Similar Species

Golden-fronted and Red-bellied woodpeckers larger, lack black-and-white face pattern. Hairy and Downy woodpeckers lack barred back.

Range

Resident from se. California east to w. Oklahoma and south through Mexico.

Black-and-white Warbler
Mniotilta varia

Sparrow-sized
Forest habitats

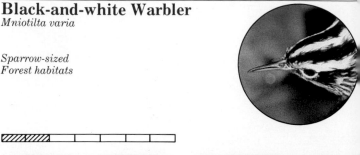

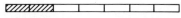

Field Marks
5¼". A small, black-and-white-striped warbler of deciduous and mixed forests; most often seen creeping around the trunks and large limbs of trees as it forages for bark insects.

Adult male (1) has black-and-white-striped crown and back, white eyebrow, black throat, and white underparts with black stripes on flanks. Female (2) and immature similar, but have whitish throats and less distinct streaks on flanks.

Song most commonly a high, thin *wee-see wee-see wee-see wee-see*; sometimes changes pitch in the middle. Note a sharp *pit* or *pit-pit*.

Similar Species
Male Blackpoll Warbler has solid black crown and white cheeks.

Range
Breeds from Northwest Territories east to Newfoundland and south to central Texas and North Carolina. Winters from s. Texas east to coastal South Carolina; also in Florida.

White-breasted Nuthatch
Sitta carolinensis

Sparrow-sized
Forest habitats; Groves

Field Marks
5¾". Slender-billed and stubby; nuthatches crawl upside down and sideways on tree trunks and branches, searching for insects in bark crevices. Common and conspicuous; frequent visitors to seed and suet feeders in winter.
Male (1, 2) has black crown and nape, gray upperparts, white face and underparts, except for brownish wash on belly; bill slim and sharp. Female similar but crown and nape gray.
Calls include a nasal *yank yank yank*. Song a short, rapid *hit hit hit hit* or *to-what what what what*.

Similar Species
Red-breasted Nuthatch has red breast and black-and-white eyeline. Brown-headed Nuthatch has black eyeline; smaller, with different calls. Brown Creeper streaked with brown above. Chickadees have black bibs.

Range
Resident from British Columbia east to New Brunswick, south to central Texas and central Florida.

Red-breasted Nuthatch
Sitta canadensis

Very Small
Boreal Forest; Eastern Deciduous
Forest; Groves

Field Marks
4½". A stubby little bird that creeps upside down or sideways on trunk and branches. More closely associated with coniferous forests than White-breasted, although more widespread in winter. Attracted to suet and seed feeders in winter.

Male (1) has black crown and line through eye, separated by white eyebrow; upperparts blue-gray; underparts bright rust. Female (2) similar but duller, with gray crown.

Call a nasal, rapid *anck-anck-anck*, like a small, tinny horn.

Similar Species
See White-breasted and Brown-headed nuthatches.

Range
Breeds from se. Alaska east to Newfoundland and south in mountains to s. California and Arizona, in Appalachians to w. North Carolina. Winters throughout most of breeding range and south to Gulf Coast and n. Florida.

Brown-headed Nuthatch
Sitta pusilla

Very Small
Southeastern Pine Forest

Field Marks
4½″. A tiny, very acrobatic forager of the pine barrens and forests of the southeastern coastal plain. Usually seen clambering up and down tree trunks and around branches, searching for insects. Unlike most nuthatches, does not form flocks with other species; rarely strays from its specialized habitat.
Adults (1, 2) dull gray above, paler gray below, with pale gray cheek patch; brown cap extends from crown to nape; bill fairly long and sharply pointed. Immatures similar to adults.
A high, piping or squeaking, double-noted call, repeated; also chatters.

Similar Species
Other nuthatches in same range somewhat more bluish-gray above. Red-breasted has bright, warm rufous-buff on breast, conspicuous white eyebrow, and distinct black line through eye. White-breasted has black crown, not brown, and rufous lower belly and undertail coverts.

Range
Resident from se. Oklahoma to s. Delaware and south to e. Texas, Gulf Coast, and central Florida.

Brown Creeper
Certhia americana

Sparrow-sized
Forest habitats; Residential Areas
and Parks

Field Marks
5½". Like a tiny brown woodpecker. Clings to vertical surfaces of tree trunks and limbs. Slender, with long, stiff tail; uses its thin, slightly curved bill to probe crevices for tiny insects. Cryptically colored and inconspicuous; creeps up trunks, often in a spiral fashion; generally starts at the base of a tree and works up.
Adults (1, 2) have brown upperparts streaked with buff; underparts off-white; thin whitish eyeline; bill very thin and somewhat downcurved.
Song short, high-pitched, thin, almost warblerlike. Call a single lisping *seep*; occasionally a *seet seet*.

Range
Breeds from se. Alaska to Newfoundland, south in mountains to Central America and w. North Carolina; s. Wisconsin east to New England. Winters south to Gulf Coast and central Florida.

Cactus Wren
Campylorhynchus brunneicapillus

Robin-sized
Southern Texas: Brushy Open
Country; Residential Areas

Field Marks
8½". A very large spotted wren, with slightly downcurved bill; much like a small thrasher. A stout, thick-legged bird of brushy country in southern Texas. Common, noisy, and conspicuous; its grating calls and chatters are common sounds in deserts and dry, brushy regions. Adults (1) have rufous crown and broad white eyebrow; upperparts grayish, streaked and spotted with brown and white; wings and tail heavily barred with black and white; underparts buff, spotted with black, heaviest on breast.
Song like a car refusing to start: a rough, low *chur-chur-chur-chur*. Various harsh scolds, chatters, and mews.

Similar Species
Thrashers slightly larger; lack streaked upperparts, barring on wings and tail.

Range
S. California east to s. Texas and south into Mexico.

Carolina Wren
Thryothorus ludovicianus

*Sparrow-sized
Forest habitats; Residential Areas
and Parks*

Field Marks
5½". A compact, energetic, warm brown bird; large for a wren. Fairly
approachable; easily lured from undergrowth by imitations of owl calls
and other bird sounds. Inhabits wooded swamps and thickets; also
frequents residential areas and parks.
Adults (1) warm reddish-brown above, paler below, with white or off-
white eyebrow and chin; flanks washed with deep warm buff; tail long,
warm reddish-brown. Immatures similar to adults.
Song a loud, repeated, whistled series of *tea-kettle, tea-kettle* or
tweeedle notes; sings all day long.

Similar Species
Bewick's Wren much grayer overall; eyebrow less distinct; tail tip
white. Winter Wren smaller; darker brown, with blackish markings
above; has buff eyestripe, not distinct eyebrow. House Wren smaller;
eastern form lacks eyebrow. Marsh Wren, rarely in same habitat, has
white streaks on back.

Range
Resident from s. Iowa east to s. New England, southward to Texas
and Florida.

Bewick's Wren
Thryomanes bewickii

Sparrow-sized
Forest habitats; Open Country;
Residential Areas and Parks

1

Field Marks
5¼". A noisy wren of brushy areas, riparian woods, and open wooded
areas with brushy understory; becoming increasingly rare in the East.
Frequently scolds and chatters as it moves about in the underbrush.
Often holds tail upright, flicks it sideways, or fans it slightly.
Adults (1) have upperparts gray-brown; whitish eyebrow; tail finely
barred with black, white on outer tips; underparts light gray; bill thin,
moderately long, and slightly downcurved.
Song reminiscent of Song Sparrow's, but thinner, beginning with two
or three high notes followed by a lower trill.

Similar Species
Carolina Wren bright rust on upperparts and buff on underparts; has
very different songs; no white in tail. Marsh Wren has white streaks
on back and shorter tail; found in marshy habitat. House Wren
smaller, without bold eyebrow, has shorter tail.

Range
Breeds locally from sw. British Columbia, southern Great Plains, and
sw. Pennsylvania south to Mexico and northern Gulf Coast. Resident
in West; in East, winters in southern part of range south to Gulf of
Mexico and Florida.

Marsh Wren
Cistothorus palustris

Sparrow-sized
Freshwater Marshes; Salt
Marshes

Field Marks

5″. An energetic and curious little wren, and the only wren likely to be found deep in a cattail, bulrush, or tule marsh. Often hidden in dense vegetation, but highly vocal and thus conspicuous.
Adults (1) have brown crown with prominent white eyebrow; upper back streaked with black and white; rump and tail rusty, the latter barred with black; underparts whitish with cinnamon-buff wash on flanks and undertail coverts. Juveniles duller, often lacking prominent white streaks on back.
Song one or two sharp guttural notes, followed by a high note, and then a rapid, raspy, but liquid trill. Calls include rattles and trill alone; also a sharp *tsuk* alarm note.

Similar Species

Sedge Wren smaller and paler, with thin white streaking on crown (solid brown in Marsh Wren); lacks bold white eyebrow; bill and tail shorter; inhabits meadows and sedge rather than cattail marshes. Carolina Wren has bright rufous back. Bewick's Wren lacks back stripes, has white tail tip; habitat different.

Range

Breeds from s. British Columbia east to Nova Scotia and south to s. California and Texas, and along Atlantic Coast to Florida; absent in interior Southeast. Winters from Washington south to Mexico, and New Jersey south to Florida; rarely in s. New England.

Sedge Wren
Cistothorus platensis

Very Small
Freshwater Marshes; Open
Country and Grasslands

Field Marks
4½". A very small, comparatively short-billed, and shy inhabitant of freshwater marshes with low, rather sparse vegetation; also occurs in wet areas of open country and grasslands. Can be difficult to observe in vegetation, but is sometimes seen flying swiftly out of cover when flushed. Like most wrens, often cocks its tail up over the back as it sings from a weed stalk or branch.
Adults (1) warm, pale brown overall, with buff-yellow wash on flanks; crown streaked with darker brown; face plain brown; in good light, white streaks on back and dark brown barring on wings may be evident. Immatures similar to adults.
Gives a harsh, two-noted song, followed by a chattering trill; often sings at night. Call a high, rich *tsip* or *chip*.

Similar Species
Marsh Wren larger, darker overall, with more prominent white on back and much longer bill. See House Wren. Other wrens not found in same habitat.

Range
Breeds from s. Saskatchewan east to s. Quebec and south to e. Kansas and w. Pennsylvania; also on Delmarva Peninsula. Winters from ne. Mexico along Gulf and Atlantic coasts to s. Virginia.

House Wren
Troglodytes aedon

Sparrow-sized
Forest habitats; Open Country;
Residential Areas and Parks

Field Marks

4¾". A common, plain little wren of brushy undergrowth. Variety of
habitats: frequents thickets and brush piles, especially around fallen
logs; in suburban areas, it readily uses nest boxes. Energetic and
spry; often holds short tail cocked at an upward angle as it moves
constantly through the brush.
Adults (1) plain, nondescript; rather rotund with a short, thin bill and
relatively short tail; brownish-gray, paler on underparts and on faint
eyebrow; wings, tail, and lower belly finely barred with blackish.
Song a rapid, liquid, bubbling chatter, often given through entire day;
also a rapid churring note and rough, buzzy scolding chatter.

Similar Species

Winter Wren smaller with very short tail; more boldly barred on
belly, different song and calls. Sedge Wren smaller, paler; has striped
crown and back; different habitat.

Range

Breeds from central British Columbia east to New Brunswick, south
to se. Arizona and n. Georgia. Winters in the Southwest, along Gulf
Coast, and north to Virginia.

Winter Wren
Troglodytes troglodytes

Very Small
Boreal Forest; Eastern Deciduous
Forest

Field Marks
4″. A stubby little wren, often vocal but usually invisible. Inhabits
dense, dank undergrowth, especially along streams and bogs in
coniferous forests and on steep slopes; less restricted in winter. Moves
mouselike underneath dense, low vegetation. When visible, it often
bounces on stubby little legs, especially when calling.
Adults (1) tiny with very short tail; dark brown upperparts and pale
brown underparts, barred with blackish on belly, wings, and tail; bill
short, thin; eyebrow inconspicuous.
Song a loud, tumbling cascade of tinkling notes and metallic
warblings. Call a vigorous *chimp-chimp*, with a quality much like a
Song Sparrow's call note, but sharper.

Similar Species
See House Wren.

Range
Breeds from s. Alaska east to Newfoundland, south to central
California, Idaho, and Great Lakes region, in mountains to n. Georgia.
Winters in s. California and Arizona, along Gulf Coast, and in the
Southeast.

Canyon Wren
Catherpes mexicanus

Sparrow-sized
Texas: Gorges and Rocky Slopes

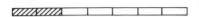

1

Field Marks
5¾". Characterized by loud, descending song that rings from sheer
cliff faces or boulder-strewn areas, especially near water. Actively
forages in rocky cracks and crannies, continuously and vigorously
bobbing up and down.
Adults (1) have warm brown upperparts, grayer on head; breast
white; belly rusty; bill rather long, slightly downcurved.
Song a loud, clear, ringing series of descending *tee-you* notes. Call a
harsh, metallic *jeet.*

Similar Species
Rock Wren much paler, without sharp contrast between breast and
belly; song different.

Range
Resident from British Columbia east to the Rocky Mountains and w.
South Dakota, south to California and w. Texas. Also Mexico and
Central America.

Rock Wren
Salpinctes obsoletus

Sparrow-sized
Texas: Gorges and Rocky Slopes

Field Marks
6″. A fairly common species of open, rocky outcroppings, talus slopes, and rocky washes in dry regions. Energetically searches rocks and crevices for food, constantly bobbing up and down. As bird flits from rock to rock, the cinnamon rump and buff tail tips flash. Often heard before seen.

Adults (1) have pale grayish-brown upperparts, lightly speckled with white; rump cinnamon; tail brown, barred with black, tipped with buff; eyebrow pale; breast grayish-white, lightly streaked, shading to buff on the lower belly and flanks.

Song usually comprises twangy couplets like *tra-lee tra-lee tra-lee tra-lee*. Call a dry, buzzy, staccato *ti-keer*.

Similar Species
See Canyon Wren.

Range
Breeds from central British Columbia east to North Dakota and south to California and central Texas. Winters in the Southwest and in South America.

Gray Catbird
Dumetella carolinensis

Robin-sized
Thickets; Residential Areas and
Parks

Field Marks
8½". A common bird found in dense thickets at the edges of
woodlands, along streams and meadows, and in parks and gardens.
Often difficult to see as it skulks deep in dense tangles; but makes its
presence known by whining, catlike *meow* call. Flips cocked tail
about, especially when excited and calling.
Adults (1) gray with black crown and long black tail; have chestnut
undertail coverts; bill slim, rather short, and blackish.
Song highly variable but consists of several phrases, some musical,
some scolding, harsh, or chattering. Phrases often separated by short
pauses, not repeated as in Northern Mockingbird or thrashers. Call a
catlike, scolding *meow*; also other harsh scold notes.

Similar Species
Northern Mockingbird paler with white in wings and tail; lacks black
crown, chestnut undertail coverts.

Range
Breeds from s. British Columbia east to Nova Scotia, south to e.
Oregon, e. Arizona, n. Texas, and Georgia. Winters from se. Texas
and se. Virginia to Central America; sparingly from Massachusetts.

Northern Mockingbird
Mimus polyglottos

Robin-sized
Thickets; Residential Areas and
Parks

Field Marks
10″. A common, active, conspicuous bird of open, shrubby places, especially in urban and suburban areas. Sings its loud, clear, complicated songs almost year-round, even at night in spring and summer. Perches conspicuously on outer branches of bushes and trees, fences, telephone poles, and antennas. Bold black-and-white wings and tail flash in flight.
Adults (1, 2) have medium gray upperparts, whitish-gray underparts; wings and tail black with white patches visible during flight; two white wing bars; bill slim and slightly downcurved.
Song rich, loud, and highly variable. Phrases vary from clear whistled warbles to harsh guttural sounds; often include imitations of other bird species, but also neighborhood sounds like squeaky clotheslines, car horns, and barking dogs. Phrases usually repeated several times before being changed. Call a loud *chak!* note.

Similar Species
Shrikes have hooked, deeper bills and black masks.

Range
Resident from n. California east to Maine and south to Mexico and Florida.

Northern Waterthrush
Seiurus noveboracensis

Sparrow-sized
Northern Wooded Swamps; Other
forest habitats

1

Field Marks
6″. Walks deliberately on the ground under vegetation, or along low branches, constantly bobbing its tail. Often hard to see, but detected by clear song or loud, metallic call note. Common in bogs and northern swamps in summer. In migration, found in any habitat near water and in dank places; occasionally dry areas in dense thickets.
Adults (1) have brownish-olive upperparts; eyestripe yellowish (occasionally whitish), tapering to a point behind eye; underparts whitish with dark spotting on throat (usually) and dark streaking on breast and sides; bill rather heavy and moderately long for a warbler; legs stocky, pinkish.
Song usually begins with three loud, clear, rising *sweet* notes, followed by a rapid, staccato, bubbling *twee-wee-wee chew chew chew*. Call a loud, ringing metallic *chink*.

Similar Species
Louisiana Waterthrush has pure white eyestripe that widens and then curves slightly upward behind eye; white throat usually unspotted; bill larger; song and call note different. Ovenbird olive-green, lacks eyestripe.

Range
Breeds from n. Alaska east to Newfoundland, south to e. Oregon and across n. United States to n. New Jersey. Winters from Mexico to South America.

Louisiana Waterthrush
Seiurus motacilla

Sparrow-sized
Streams and Brooks

Field Marks
6″. A small bird with a streaked breast; common along streams with
dense wooded borders. Favors running water over swamp or lake
margins, which it may visit on migration. Has a habit of bobbing tail
while walking at the water's edge searching for insect prey.
Adults (1) have cold grayish-brown upperparts with a prominent
white eyebrow that extends well back behind eye; underparts whitish
with a slight buff wash on flanks and bold dark streaks on breast;
flanks are obscurely streaked with brownish; dark bill thin and
pointed; legs pinkish.
Song a clear, shrill *seeur seeur seeur persee-ser.* Call a loud *chink.*

Similar Species
Northern Waterthrush usually has a yellowish wash in face and
throat; yellowish eyebrow that tapers just behind eye; flanks heavily
streaked; prefers slow-moving or still water; has a different song.

Range
Breeds from e. Nebraska east to central New England and south to e.
Texas, Louisiana, and North Carolina. Winters in West Indies,
Central and South America.

Ovenbird
Seiurus aurocapillus

Sparrow-sized
Eastern Deciduous Forest;
Thickets

Field Marks
6″. A common, rotund bird of the forest floor, especially deciduous slopes. A secretive warbler that walks on the ground, often under vegetation or on low branches, bobbing and teetering as it creeps along. May be hard to see, but the remarkably loud, ringing songs, heard everywhere in spring, indicate how common Ovenbirds actually are.

Adults (1, 2) have olive-green upperparts; orange crown bordered by black stripe; bold white eye-ring; underparts whitish, boldly streaked with black spots.

Common song a ringing *teacher teacher teacher teacher teacher*, steadily increasing in volume. Call a loud, dry, saucy *tcheek*.

Similar Species
Waterthrushes have bold eyestripes, brownish-olive upperparts.

Range
Breeds from ne. British Columbia east to Newfoundland, south to e. Colorado and n. Georgia. Winters from Gulf Coast and Florida to South America.

Hermit Thrush
Catharus guttatus

Sparrow-sized
Forest habitats; Residential Areas
and Parks

Field Marks
6¾″. A common breeder of coniferous or mixed woodlands in summer;
in winter, also found in brushy growth, including gardens, especially
those with berries. In breeding season, sings beautiful, haunting,
ethereal phrases, most often at dawn and dusk. Plumage variable, but
always distinguished by contrasting rusty tail. Like most thrushes,
usually on or near ground. Often flicks its wings.
Adults (1) have brownish-olive upperparts, with bright rusty tail; eye-
ring whitish; underparts whitish with buff on breast and grayish-buff
flanks, breast heavily spotted with blackish.
Song begins with prolonged, clear whistled note followed by a short
series of rapid, slurred, flutelike changes; phrases are separated by
pauses and frequently change in pitch. Call a distinctive, low, slightly
liquid *chuck*.

Similar Species
See other thrushes; Hermit is only thrush with contrasting rusty tail.
Fox Sparrow stouter, with conical bill; eastern birds strongly rusty.

Range
Breeds from Alaska east to Newfoundland, south to s. California,
Great Lakes region, and Maryland. Winters from Washington and
s. New England south.

Swainson's Thrush
Catharus ustulatus

Sparrow-sized
Boreal Forest; Eastern Deciduous
Forest; Residential Areas and
Parks

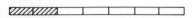

Field Marks
7″. A brownish thrush, best identified by conspicuous buff spectacles
and breast. Found in our range chiefly in summer, in coniferous and
mixed woods and tall deciduous shrubbery; in migration, also in
residential areas and city parks.
Adults (1) have olive-brown upperparts; spectacles and breast
strongly buff; dark but diffuse spotting on throat and breast; flanks
buff-brown.
Song an upward-spiraling, slurred series of harmonic, flutelike notes.
Call a liquid *whoit* or *whoit-whoit*.

Similar Species
Other thrushes lack strong buff spectacles.

Range
Breeds from central Alaska east to Newfoundland, south to s.
California and New Mexico; in East, to Great Lakes region and
sparingly in mountains to West Virginia. Winters in Central and
South America.

Gray-cheeked Thrush
Catharus minimus

Sparrow-sized
*Boreal Forest; Eastern Deciduous
Forest; Residential Areas and
Parks*

Field Marks
7¼". Breeds in boreal coniferous forests of Canada and New England
mountains, especially in dense spruce and spruce-fir near treeline. In
migration, found in a variety of brushy thickets. Generally rather
quiet and shy, it is probably the least conspicuous thrush.
Adults (1) have dark gray-brown to brownish-olive upperparts; tail
sometimes more olive or brown than back, especially in fall; cheeks
very clear, medium gray; hint of light gray eye-ring; underparts
whitish (sometimes with buff wash on breast), heavily spotted with
dark gray; sides and flanks washed with gray.
Song a liquid but somewhat nasal, short series of descending, rolling,
slurred notes, reminiscent of Veery. Calls include a loud *vee-ah*,
similar to Veery's; also a dry, loud *what*.

Similar Species
No other thrush has such a clear, cold-gray cheek. Swainson's has buff
eye-ring; Hermit has rusty tail; Veery more rufous above, with fainter
breast-spotting.

Range
N. Alaska across central Canada to Newfoundland and south to
British Columbia and se. New York. Winters in West Indies, Central
and South America.

Wood Thrush
Hylocichla mustelina

Robin-sized
Eastern Deciduous Forest; Other
forest habitats; Residential Areas
and Parks

Field Marks
7¾". A plump, rufous thrush, abundant in residential areas, deciduous woods, and thickets. Forages on the ground where it hops and scratches in the leaf litter while searching for worms, caterpillars, insects, and ants. Effortless, languid, flutelike song distinctive. Adults (1) have russet to cinnamon-brown upperparts; face has a thin white eye-ring and narrow black-and-white streaks on the cheek. Bold, triangular or roundish black spots cover the white underparts; bill dark, with the basal two-thirds of lower mandible flesh-colored; legs are also flesh-colored.
Song flutelike, liquid, and varied; includes characteristic *whee-do-lay* phrase in which the last note is usually trilled. Call a low *qoit*; alarm note a sharp *pit*, repeated several times.

Similar Species
Hermit Thrush is reddish only on the tail; has smaller spots restricted to the breast; typically cocks up tail. Veery has only light spotting on throat; shape smaller and slimmer; only a trace of an eye-ring.

Range
Breeds from se. South Dakota east to s. Maine, south to se. Texas and n. Florida. Winters in Central America.

Veery
Catharus fuscescens

Sparrow-sized
Eastern Deciduous Forest;
Residential Areas and Parks

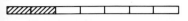

Field Marks
7″. Best known for its beautiful, haunting song. A brownish thrush of
a variety of forest habitats, it spends most of its time on the ground,
foraging for insects and fallen fruit. Can be difficult to observe in
dense cover, but imitations of birdcalls will sometimes induce it to
venture out.
Adults (1) cinnamon-brown above, buff below, with light brown spots;
sides of throat also buff; flanks pale gray. Immatures similar but
darker, with spots on head and back.
Song a clear, mellow *vee-ur, veer, veer*, with a descending, flutelike
quality. Call a soft, descending *whew*.

Similar Species
Other thrushes lack uniform cinnamon-brown upperparts; have
heavier spotting below.

Range
Breeds from s. British Columbia east to central Newfoundland and
south to Great Lakes region, in mountains to New Mexico and
Georgia. Winters in Central and South America.

Brown Thrasher
Toxostoma rufum

Robin-sized
Thickets; Residential Areas and
Parks

Field Marks
11½″. A sleek, long-tailed bird with rufous upperparts. Found in
thickets, vine tangles, woodland edges, and gardens; most common in
the southeastern United States. Rather secretive and solitary.
Adults (1) have dull cinnamon-brown to rufous upperparts, brightest
on the rump and upper-tail coverts; face and forehead grayish-buff;
wings have two thin white wing bars; underparts buff-white with
numerous elongated, blackish-brown spots; eyes orange or yellow; bill
slightly downcurved, dark, with pale base to the lower mandible.
Call is a loud, sharp *tsick!* Song much like a Gray Catbird's but richer,
more melodious, and sung in paired phrases.

Similar Species
Wood Thrush has streaked face and roundish spots below. Long-billed
Thrasher (Texas only) has a darker face; darker gray-brown
upperparts, brightest on tail, all-black bill; and blacker spots on
whiter underparts.

Range
Breeds from se. Alberta east to s. Maine, south to Colorado, e. Texas,
and Florida. Winters from Texas eastward as far north as Illinois, and
along Atlantic Coast as far north as Massachusetts; rarely to Pacific
Coast.

Long-billed Thrasher
Toxostoma longirostre

Robin-sized
Southern Texas: Streamside
Forest; Thickets

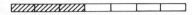

Field Marks
11½". A secretive, ground-foraging bird of willow, mesquite, and
acacia bottomland; within the United States, restricted to southern
Texas. Like other thrashers, searches for food by using its bill to rake
through leaf litter.
In adults (1), the head, back, and wings are dark brown, the tail is a
somewhat brighter russet or cinnamon-brown; face dark gray;
underparts whitish with long, black spots and streaks; bill entirely
black.
Calls similar to Brown Thrasher's, including low, sharp *tsick!* Song
also similar, with a jumble of repeated melodious phrases.

Similar Species
Brown Thrasher has paler grayish-buff face, brighter rufous
upperparts, and pale base to lower mandible; underparts buffier,
with more brownish spots.

Range
Resident in s. Texas. Also in e. Mexico.

Curve-billed Thrasher
Toxostoma curvirostre

Robin-sized
Texas and Oklahoma: Thickets;
Brushy Open Country

Field Marks
11″. A common, large, long-billed grayish bird of deserts, thorn scrub, suburbs, brushy areas, and thickets; in the East, found in suitable habitats in Texas and Oklahoma. Often seen perched on a rock or cactus; flies rapidly, in a low, direct fashion. More conspicuous than most other thrashers.
Adults (1, 2) pale gray-brown above; underparts slighly paler gray-brown, with some dark spots; bill long and strongly downcurved; eyes yellow; tail long, with white outer corners that flash in flight. Some birds in eastern part of range have white wing bars; western birds may lack white tail corners. Immatures have shorter, less downcurved bill, less spotting on underparts; tail shorter.
Call a sharp, liquid *whit-wheet*; song a melodic trill with some repeated phrases.

Similar Species
See Sage Thrasher.

Range
Breeds from central Arizona east to nw. Oklahoma and south to Mexico and w. Texas. Winters from s. Arizona south.

Sage Thrasher
Oreoscoptes montanus

Robin-sized
Brushy Open Country

Field Marks
8½". A common thrasher of open countryside with brushy growth;
fairly abundant in plains and deserts of the West. Feeds on berries
and other fruits, as well as on insects. Often flicks its tail like a
Northern Mockingbird; when alarmed, scurries quickly into brushy
cover.
Adults (1, 2) gray-brown above, paler below with dark black streaks;
two white wing bars (sometimes indistinct); tail fairly long, with white
outer corners. Immatures darker brown above, with less well-defined
streaks below. In all plumages, bill fairly straight.
Song a warbling series of mockingbirdlike notes.

Similar Species
Curve-billed Thrasher larger, with much longer, downcurved bill;
lacks streaks below.

Range
Breeds from s. British Columbia east to s. Montana, south to e.
California and w. Texas. Winters from s. Arizona to w. Texas
and south.

Sprague's Pipit
Anthus spragueii

Sparrow-sized
Open Country and Grasslands

1

Field Marks
6½″. A sparrowlike, short-tailed pipit, common in short-grass prairies
in summer and low, dry, grassy fields in winter. Tends to crouch
when alarmed, taking flight at the last moment, often uttering its
characteristic squeaky call. Walks without pumping tail. Sings while
circling high in the air.
Adults (1) are cryptically colored, brownish birds with streaked crown
and back, which in fall looks scaled; dark eyes are prominent in plain,
pale buff face; eyebrow slightly paler; sides of throat and upper breast
have fine dark streaks; underparts otherwise whitish; tail black with
pale edgings and white outer feathers; bill pale; legs pinkish.
Call a loud, harsh *tsweet*. Song a descending series of grating,
thrushlike whistles.

Similar Species
Water Pipit has patterned face, pumps tail, and has gray legs. Vesper
Sparrow and female Smith's and Lapland longspurs have boldly
marked faces.

Range
Breeds from Alberta and Manitoba south to Montana and Minnesota.
Winters from southern Great Plains south to Mexico.

Water Pipit
Anthus spinoletta

Sparrow-sized
Open Country and Grasslands

Field Marks
6½″. Sparrow-sized, but slimmer and longer with a thinner bill.
Highly terrestrial; walks on ground with constant bobbing motion and
tail-wagging. Rarely perches above ground level. Found in alpine
tundra (in summer), grasslands, agricultural areas, and barren
ground. Often gives diagnostic calls in flight and on the ground. White
outer tail feathers flash conspicuously during undulating flight. Often
highly gregarious.
Breeding adults have grayish-brown upperparts with very light dusky
streaking on head and back; buff eyebrow; underparts pinkish-buff
with light streaking on breast and sides. Nonbreeding adults (1) and
immatures similar, but with upperparts darker olive; underparts more
heavily streaked, pinkish cast absent.
Song a rapid musical series of *che* and *cheedle* notes, given in flight.
Call a sharp *pip*, *pip-it* or *pi-pi-pi-pit*.

Similar Species
Sprague's Pipit has paler, more streaked upperparts, finer streaking
on breast, pale legs, and different call: a loud *sweep*.

Range
Breeds from Alaska east across n. Canada to Newfoundland and south
in mountains to New Mexico and n. Maine. Winters along Pacific
Coast and across s. United States, rarely north to New York.

McCown's Longspur
Calcarius mccownii

Sparrow-sized
Open Country and Grasslands

Field Marks
6″. A stout longspur, best identified by tail pattern, large bill, and
call notes. Found in short-grass prairies, plowed fields, and barren
areas. Cryptic and secretive when not singing. Usually flushed from
ground before being seen. Flight undulating.

All plumages have white tail with inverted black T in middle; bill large
and pinkish (but black in breeding male); wing tips almost reach
end of short tail. Adult male (2) in summer has brownish streaked
upperparts; head and underparts gray, with black crown, mustache,
and crescent on breast; rusty patch on shoulder. Winter male, female
(1), and juvenile have buff to whitish underparts without streaks.
Winter males often retain grayish cast, especially on rump, and trace
of crescent on breast.

Song a series of twittering warbles delivered in flight. Flight call a
dry rattle; also a unique, popping *poik*.

Similar Species
Tail patterns of other longspurs different; tails longer, bills smaller,
underparts streaked; lack *poik* call.

Range
Breeds from s. Alberta east to North Dakota and south to Colorado.
Winters from Arizona east to Oklahoma and central Texas, south to
Mexico. Occasionally wanders to California and Atlantic Coast.

Dickcissel
Spiza americana

Sparrow-sized
Brushy Open Country

Field Marks

6½″. A streaked bunting of grassy, weedy, brushy, and agricultural
areas of the Midwest. Feeds on or near ground in vegetation, but is
conspicuous on fences and tall brush, especially when singing.
Generally gregarious.
Breeding male (2) has bright yellow eyestripe, mustache, and breast;
V-shaped black bib on breast; head and rest of underparts mostly
grayish; upperparts brown, streaked with black; rufous shoulder
patch. Female (1) similar but duller, lacking black bib. Winter male
like breeding male but duller. Immature (3) plain, with traces of adult
face pattern; breast lightly streaked.
Song two sharp notes followed by buzzy notes: *dick dick dickcissel*.
Call a short buzzy note, often heard at night during migration.

Similar Species

Females and immatures resemble female House Sparrow, but the
latter lacks yellow on breast and distinctive mustache. Meadowlarks
much larger with long bills, white in tail. Similar Bobolink plumages
more buff, with crown stripe; lack mustache.

Range

Breeds from e. Montana east to central Ohio and south to Texas and
central Gulf Coast states. Winters from Mexico to South America.

Lapland Longspur
Calcarius lapponicus

Sparrow-sized
Open Country and Grasslands;
Beaches and Coastal Dunes

Field Marks

6½″. Breeds in northern arctic tundra. In winter found in plowed fields, grassy areas, shores, and beaches. Often mixes with Horned Larks and Snow Buntings. Gives distinctive calls when flushed from ground. Tail pattern unique for a longspur. Flight style deeply undulating.

All plumages have diagnostic dark tail with white outer tail feathers. Breeding male (2) has black head and breast broken by white line extending from eye back and down around sides of breast; nape chestnut; upperparts streaked and blotched with brown; wings brown with rufous. Female (1), winter male, and immature similar to breeding male but duller, with conspicuous ear patch outlined by black, and dark mustache; underparts variably streaked.

Song a rapid, musical warble, delivered in flight. Calls a dry three-noted rattle and a diagnostic, descending, whistled *tew* note.

Similar Species

All other longspurs have more white in tail; lack *tew* note; not as dark overall.

Range

Breeds from n. Alaska across n. Canada. Winters from California east to s. Maine and south to central Texas.

342

Chestnut-collared Longspur
Calcarius ornatus

Sparrow-sized
Open Country and Grasslands

Field Marks
6″. Breeds in grassy prairie of upper Midwest. In winter, found in tall-grass areas more than other longspurs (where it is often invisible until flushed), also in short grass and plowed fields. When flushed, tail pattern and flight call identify this species. Gregarious in winter. Flight deeply undulating.

All plumages have white tail, black triangle at base. Breeding male (2) has black, white, and buff head; nape chestnut; upper two-thirds of underparts black; belly white; upperparts brown, streaked with black and buff. Adult female (1), winter male, and immature duller, lack distinctive head pattern of breeding male; may show trace of black breast; generally relatively buffy with streaked crown.

Song a rapid warble usually given in flight. Call a very distinctive musical *deedup, deedle, deedleup*; occasional dry rattle.

Similar Species
No other longspurs have distinctive *deedup* notes or distinctive tail pattern.

Range
Breeds from se. Alberta east to w. Minnesota, south to ne. Colorado and n. Nebraska. Winters from Arizona east to e. Texas and south to Mexico.

Horned Lark
Eremophila alpestris

Sparrow-sized
Open Country and Grasslands

Field Marks
7¼". A common terrestrial species found in open country, including grasslands, fields, dunes, and extensive lawns. Often seen in large flocks, walking energetically across the landscape searching for food. Look for other field birds such as longspurs in flocks of Horned Larks. As with most birds of open country, the call note is diagnostic. The black tail flashes white outer borders in flight.

Highly variable, but face pattern distinctive. Adult (1, 2) has black forehead with extended "horns," black line from bill to eye curving down behind the eye, and black breast shield; upperparts pinkish-brown; ground color of head and underparts varies from pure white to bright yellow. Juvenile has back and head dark brown, covered with white spots.

Call a clear *tsee-ee, tsee-titi.* Song a weak but carrying series of high twittering and tinkling notes.

Similar Species
Other field species lack striking face pattern, have different calls.

Range
Breeds and winters from Alaska east to Newfoundland and south throughout most of United States to Mexico; rarely in southeastern coastal states.

344

Smith's Longspur
Calcarius pictus

Sparrow-sized
Open Country and Grasslands

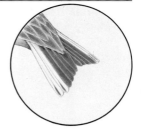

Field Marks
6¼″. An uncommon longspur that often goes undetected, especially in winter, because of its cryptically patterned plumage and skulking, ground-feeding habits. Breeds on drier portions of tundra and winters in dry short-grass fields. Males sing on the ground.
Breeding male (1) a rich buff on underparts, chin, throat, and hindneck; head black with white eyebrow and spot on cheek; tail blackish, with outer three feathers partly white, the outermost largely white. Female (2) brown above with dark streaking, pale median crown line, buff eyebrow, and dark frame around ear patch; underparts cinnamon-buff with fine, dark streaks on sides and flanks. Both male and female have a whitish shoulder patch, similar tail patterns, and dark bill with pale lower mandible. Immatures and winter adults are like adult female.
Call a dry, rapid rattle of clicks. Song a sweet warble combined with clicks, a high *tew*, concluded with a quick *wee-chew*.

Similar Species
See other longspurs and Vesper Sparrow.

Range
Breeds locally from central Alaska east to n. Ontario. Winters from Nebraska south to e. Texas; occasionally wanders farther east.

Western Meadowlark
Sturnella neglecta

Robin-sized
Open Country and Grasslands

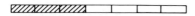

Field Marks
9½". Like the Eastern Meadowlark, this is a common stocky bird of
open country. The striking black V on the bright yellow breast
identifies it at great distance as a meadowlark, and in most of the
West, as a Western Meadowlark. In the Southwest and Midwest,
however, one must hear the bird or get a very close look to separate
it from the Eastern Meadowlark.

All plumages have long, pointed bill. Adults (1, 2) have head striped
with black and white; cheeks yellow; lores, throat, and underparts
mostly bright yellow, with striking black V on breast; upperparts
streaked and blotched with browns, buff, and white; tail short with
conspicuous white outer portions.

Song has several clear, whistled introductory notes followed by a
rapid series of cascading, bubbling, rich flutelike notes. Call a mellow
chuck; also a buzzy rattle in flight.

Similar Species
Eastern Meadowlark has clear, whistled song, without bubbling
quality; lacks *chuck* note and instead has unique twangy call. Cheek
white in Eastern, yellow in Western.

Range
Breeds from British Columbia east to w. New York and south to
California, Great Lakes region, and central Texas. Northern birds
move south in winter.

346

Eastern Meadowlark
Sturnella magna

Robin-sized
Open Country and Grasslands

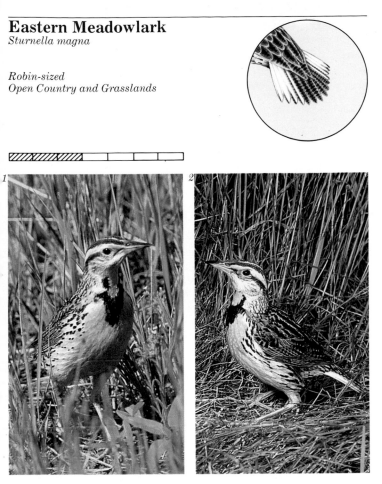

Field Marks
9½". Meadowlarks are common, short-tailed, long-billed birds of
meadows and fields. Both species show a striking black V on yellow
breast and conspicuous white outer tail feathers. Flight style
characteristic; consists of series of rapid, stiff wingbeats broken by
brief glides with wings held rigidly downward. Easily seen from
roads, where it flushes before approaching cars, or singing from posts
or fences, where it perches on stocky legs. Two species best separated
by voice.
All plumages have long, pointed bill. Adults (1, 2) have head striped
with black and white; cheek white, with no yellow; lores, throat, and
underparts mostly bright yellow, with striking black V on breast;
upperparts streaked and blotched with browns, buff, and white; tail
short with obvious white outer feathers.
Song has several clear, whistled, descending notes: *see-you, see-yeer*.
Diagnostic call a twangy, buzzy *dzert*; also gives buzzy rattle in flight.

Similar Species
Western Meadowlark's song and call different; has yellow on cheek.

Range
Breeds from Minnesota east to central Maine and south to Arizona
and Florida. Northern birds move south in winter.

Altamira Oriole
Icterus gularis

Robin-sized
Southern Texas: Streamside
Forest

Field Marks
10″. A large, bright orange oriole, fairly common in the Rio Grande valley of southern Texas. A tropical bird that builds a stockinglike woven nest nearly two feet long. Flight is somewhat jerky and typically short.

Adults (1, 2) have head, rump, and underparts brilliant orange—tending to reddish on head and yellowish on middle of belly; throat, lores, back, wings, and tail are black; shoulder orange, lower wing bar white; has small white patch at the base of primaries and some white edgings on wing; bill deep, short, with a perfectly straight ridge on upper mandible. Winter adults have dull orange edges on the back feathers. Immatures (3) are dull yellow where the adult is orange. Juveniles are yellow, mostly lack black on throat, and have brownish-olive back, wings, and tail.

Call a low, rasping *ik-ik-ik*. Song a disjointed series of whistles.

Similar Species
Hooded Oriole smaller, slimmer, with a slightly downcurved, thinner bill, more black between eye and bib, and two white wing bars.

Range
Resident in s. Texas. Also in Mexico and Central America.

348

Spot-breasted Oriole
Icterus pectoralis

Robin-sized
Southern Florida: Residential
Areas and Parks; Forest habitats

Field Marks
9½". A large black-and-orange oriole, introduced to southeastern
Florida from Central America. Uncommonly encountered in
residential parks and gardens; the only oriole in southern Florida
during summer.
Adults (2, 3) dark orange with a black throat, black spots on breast,
and black lores, back, wings, and tail; shoulder orange; wings have a
small white patch at base of primaries and a large white patch on
secondaries. Immatures (1) similar but duller; may lack spots on
breast. Juveniles are yellowish with olive back, wings and tail; throat
mottled with black.
Call a harsh rattle. Song a long series of flutelike whistles.

Similar Species
Male Northern "Baltimore" Oriole has black head and orange in tail.
Altamira Oriole of Texas has straight upper mandible, lacks black
breast spots.

Range
Established on southeastern coast of Florida; more abundant in
Central America.

Northern Oriole
Icterus galbula

Robin-sized
Eastern Deciduous Forest; Other
forest habitats; Residential Areas
and Parks

Field Marks
8¾″. Although brightly colored, the "Baltimore" Oriole is most often
first detected by a loud, clear song and insistent chatters coming from
the interior of a tall deciduous tree. When seen, the bold, bright
pattern of the male is distinctive. A common summer bird of gardens,
parks, and open woodlands.

Adult male (1, 2) has black head and back; brilliant, deep orange
underparts, shoulder, and rump; wings black with white wing bar; tail
black with orange outer patches toward tip. Adult female (3) has
olive-gray upperparts and yellowish-orange underparts, strongest on
breast and undertail coverts; two whitish wing bars; sometimes shows
dark hood. Immature male like adult female, but usually brighter with
more dark on hood; immature female duller.

Song a loud, clear series of warbled and whistled notes; variable. Call
a *keep-oh*, the second syllable lower; also a rapid chatter.

Similar Species
Adult male Orchard Oriole burnt orange, much darker than Northern;
females and immature Orchard much smaller, with shorter bill; tail
shorter; greener color; *chuck* note. See Hooded Oriole.

Range
Breeds from central Alberta east to Nova Scotia and south to central
Louisiana; rare along Atlantic Coast south of Virginia. Winters from
Mexico to South America, sparingly along Gulf Coast.

Hooded Oriole
Icterus cucullatus

Robin-sized
Southern Texas: Streamside
Forest; Residential Areas and
Parks

Field Marks
8″. A slim oriole with a thin, slightly downcurved bill. Common and
widespread in riparian woodlands, desert oases, and towns; often
associated with palm trees. Conspicuous and noisy, emitting constant
chattering and frequent *wheenk* notes.

Adult male (1, 2) bright yellow-orange, with black bib extending from
eye to upper breast; back and tail black; rump yellow-orange; wings
black with two white bars. Females (3) greenish-yellow, brightest on
the underparts; two white wing bars. Winter adult male has back
lightly barred with buff. Immature male like adult male, but orange-
yellow replaced by yellow-green.

Song a rapid series of warbles. Call a unique nasal *wheenk* or *wheet*;
also a soft, dry chatter.

Similar Species
Altamira Oriole much larger, with orange shoulder, one white wing
bar. Male "Baltimore" Oriole has black hood; female has deeper,
straight bill; orange cast on underparts. Female and immature
Orchard Oriole quite similar, but smaller with very short tail and
shorter, straight bill; call different.

Range
Breeds from n. California, where local, east to s. Utah and south to s.
Texas. Winters in Mexico; rarely in Southwest.

Orchard Oriole
Icterus spurius

Sparrow-sized
Open Country; Residential Areas;
Streamside Forest

Field Marks
7¼″. A small, dark, short-tailed oriole, common in lowlands with scattered trees and along rivers. Feeds almost exclusively on insects. It builds a shallow, woven nest. Male typically sings from a high perch, occasionally flying upward and singing while dropping back to perch.
Males (2, 3) have black head, throat, back, wings, and tail; rump and underparts deep chestnut. Females (1) yellow with a greenish back and tail, and two narrow white wing bars. Immature males like adult females, but with black face and throat. Bill short and straight in all plumages.
Calls include a weak, grating rattle and characteristic *chuck*. Song a musical, varied series of whistles.

Similar Species
Female and immature Hooded Oriole have slightly downcurved bill, longish tail, and different call. See Northern Oriole.

Range
Breeds from s. Manitoba east to s. New England, south to e. New Mexico and n. Florida. Winters from Mexico to South America; occasionally in California.

American Robin
Turdus migratorius

Robin-sized
Residential Areas and Parks;
Urban Areas; Forest habitats;
Open Country

Field Marks
10″. A familiar, large, heavy-bodied thrush of parks and gardens, as well as wooded and open areas. Often seen running along lawns, and stopping and cocking head to side while searching for food. During nonbreeding season, often seen in large flocks, especially in evening at huge communal roosts.
Adult male (1, 3) has dark gray-brown upperparts, tending toward black on head; underparts bright brick-red, except throat, which is whitish with dark streaks, and lower belly, which is white; bill yellow. Adult female (2) similar but duller. Juvenile has same shape and basic color pattern, but is heavily spotted on underparts.
Song a rich, fluid series of distinct phrases that rise and fall. Calls include soft *tut . . . tut . . . tut* and loud *wink*.

Similar Species
Varied Thrush, *Ixoreus naevius*, vagrant in East, has orange eyestripe and wing bars, dark breast band.

Range
Breeds from Alaska east to Newfoundland, south to California, e. Texas, and South Carolina; occasionally along Gulf Coast. Winters from British Columbia to Newfoundland, south to California, s. Texas, and Florida.

Yellow-headed Blackbird
Xanthocephalus xanthocephalus

Robin-sized
Freshwater Marshes; Open
Country and Grasslands

Field Marks
9½". A common, loud, conspicuous blackbird of freshwater marshes
and nearby farmlands. Often detected by harsh, loud rasping and
squeaking calls coming from dense reeds. Gregarious; often seen with
other species of blackbirds.
Adult males (1) have black body, white wing patch; entire head and
upper breast rich yellow except for black around eye. Female (2)
brownish with dull yellow breast and throat. Juvenile similar to
female but paler.
Song a loud, raspy note followed by a long sound like a squeaky hinge.
Call a low, harsh *cruck*.

Similar Species
Male Bobolink has white in wings and yellow on back of head, but
rump white, breast black. See Horned Lark.

Range
Breeds from central British Columbia east to Wisconsin, south to
s. California and w. Illinois. Winters from California to w. Texas,
occasionally to Florida and along Atlantic Coast.

Audubon's Oriole
Icterus graduacauda

Robin-sized
Southern Texas: Streamside
Forest

Field Marks
9½". A very secretive, large oriole, uncommon in dense tangles and
woods of the lower Rio Grande valley in Texas. Its whistled song has
a human quality; may sing at any time of year from a hidden perch
deep in foliage.
Adults (1) have a yellow body with a greenish back; head, wings, and
tail are black, with white edges to wing coverts; shoulder yellow;
lower portion of hood streaked with yellow. Females somewhat duller
than males. Immatures duller with grayish-brown wings and tail.
Juveniles lack black hood.
Call resembles Altamira Oriole's grating *ik-ik-ik*. Song a slow, mellow
series of whistles.

Similar Species
Juvenile Altamira Oriole has deep, short, triangular bill and olive-
colored wings and tail.

Range
Resident along Rio Grande in s. Texas. Also in Mexico.

Bobolink
Dolichonyx oryzivorus

Sparrow-sized
Open Country and Grasslands;
Freshwater Marshes; Salt
Marshes

Field Marks
7″. Rather slim with a conical bill; tail feathers sharply pointed.
Common in most parts of its range; a conspicuous bird of fields and
meadows. Males noisy, singing bubbly songs in flight on stiffly held
wings with erratic, shallow wingbeats. Upon arrival in spring, the
fields can be packed with Bobolinks chasing each other back and forth
and singing.
Adult males (1) black with yellow-buff to buff-white hindneck, white
rump, and white shoulders. Females (2), winter males, and immatures
buff with dark streaks on back; also dark crown stripes and line
behind eye; underparts variably streaked.
Song a loud bubbling series of warbled, reedy notes. Call a distinctive
eenk.

Similar Species
Female Red-winged Blackbird much darker and much more heavily
streaked on breast. Savannah Sparrow smaller, heavily streaked on
breast, not buff. Male Lark Bunting has white only on wings.

Range
Breeds from se. British Columbia east to Nova Scotia, south to n.
California, central Great Plains, and New Jersey. Winters in South
America.

Red-winged Blackbird
Agelaius phoeniceus

Robin-sized
Freshwater Marshes; Open
Country; Salt Marshes

Field Marks
8¾". A common species of wet fields, meadows, and marshes. Males
quite conspicuous, with jet-black plumage and fiery red shoulders
frequently flared in territorial display. Highly gregarious in winter;
often found in open fields with other species of blackbirds.
Adult males (1) black, except for bright red shoulder patch bordered
on bottom by buff-yellow; red not always visible when perched.
Females (2) and immatures heavily streaked with dark gray-brown on
underparts; some show red on shoulder.
Song a loud, forced *konk-a-ree*. Calls include a low *chuck* and a
metallic *kink*.

Similar Species
Other female blackbirds not streaked. Female Bobolink strongly buff.
Juvenile Brown-headed Cowbird has streaks, but is smaller, paler,
and has conical bill.

Range
Breeds from central Alaska east to s. Newfoundland and south
throughout United States. Winters from s. British Columbia and
s. New England southward.

Rusty Blackbird
Euphagus carolinus

Robin-sized
Northern Wooded Swamps; Open
Country; Freshwater Marshes

Field Marks
9″. A medium-sized blackbird that breeds near water in spruce bogs
across northern portion of the continent. Frequently seen in small
groups with other blackbirds; prefers pool margins and other wet
areas in farmland; also woodland. Has slim, pointed, somewhat
mockingbirdlike bill.
Breeding males (2) are dull black with yellow eyes. Breeding females
dark gray with black lores and yellowish eyes. In fall and winter both
sexes (1) have extensive rusty edgings on the upperparts, prominent
buff eyebrows, and black lores. Immatures are similar to winter
adults, but paler cinnamon-buff below, browner above.
Call a harsh *shack*. Song a squeaky *kish-a-lee*, like a creaking hinge.

Similar Species
Immature male Brewer's Blackbird, *Euphagus cyanocephalus*, is
a western visitor to Great Lakes region in summer, Southeast in
winter; has greenish gloss on wings, dull brownish feather edgings,
deeper bill that is not sharply pointed; adult male has purple sheen on
head. Common Grackle much larger with long bill and tail, purple
sheen on wing.

Range
Breeds from Alaska east to Newfoundland and south to central British
Columbia and n. New England. Winters from s. North Dakota east to
New Jersey and south to Gulf Coast and Florida.

European Starling

Sturnus vulgaris

Robin-sized
Residential Areas and Parks;
Urban Areas; Open Country

Field Marks

8½". Introduced from Europe. Aggressive and adaptable; has spread rapidly throughout North America. Found almost everywhere, but especially in areas associated with man's activities. Highly gregarious; large flocks occur in fields and on lawns, often with cowbirds and blackbirds. In flight, has distinctive short, squared tail and short, wide, pointed wings; wingbeats are rapid and stiff, alternating with brief glides.

Breeding adult (2) black with metallic iridescence, some light spotting on back; long, sharp, bright yellow bill. Winter adult (1) glossy black, covered with little light speckles; bill darker. Juvenile shaped like adult but flat brownish-gray overall; bill dark.

Song a rapid, soft series of notes that include repeated phrases, whistles, squeaks, and bubbling sounds; often incorporates imitations of other species into song. Other vocalizations quite variable, but high-pitched squeaking a common call.

Similar Species

Blackbirds and cowbirds have longer tails, shorter dark bills.

Range

Resident from se. Alaska east to Newfoundland and south throughout United States.

Common Grackle
Quiscalus quiscula

*Robin-sized
Residential Areas and Parks;
Forest habitats; Open Country;
Groves*

Field Marks
12½″. A large, abundant blackbird of parks, gardens, and a variety of
open habitats, including fields and marshes. Often seen walking across
lawns on stocky legs with deliberate strides. Highly gregarious, noisy,
and conspicuous, forming large flocks.

All plumages large, with a relatively heavy, long bill and a long,
wedge-shaped tail. Adults (1, 2) black, washed with a strong metallic
sheen and a yellow eye. Juveniles (3) are flat grayish-brown with
dark eyes.

Most common call a loud *clack*; also produces various loud squeaky
notes like rusty hinges.

Similar Species
Boat-tailed and Great-tailed grackles both larger and more slender
with much longer tails; females brown, not glossy black. Rusty and
Brewer's blackbirds much smaller, with smaller bills; shorter tails
without deep wedge.

Range
Breeds from ne. British Columbia east to sw. Newfoundland and
south to central Colorado, s. Texas, and Florida. Winters from s.
Minnesota and s. New England south to e. Texas and Florida.

Boat-tailed Grackle
Quiscalus major

Pigeon-sized
Freshwater Marshes; Salt
Marshes; Open Country;
Residential Areas and Parks

Field Marks
16½″. A large, long-tailed blackbird of coastal salt marshes, except in
Florida where it also inhabits freshwater marshes and residential
parks. Quite gregarious. Male usually displays from the top of a post,
bush, or tree, giving a loud, whining call and puffing his head
feathers.
Males (2) glossy blue-black with a dark eye (Gulf Coast and Florida)
or dull yellowish eye (Atlantic Coast); tail is long and keel-shaped.
Females (4) cinnamon-brown with blackish wings and tail, and pale
buff eyebrow; eyes like the male's; tail long, but not as long as male's,
and rarely keeled.
Call a loud, ringing *weeh-weeh-weeh*; also harsh *check* notes and
whistles.

Similar Species
Great-tailed Grackle, *Quiscalus mexicanus* (1, 3), of coastal Texas
and Louisiana, has clear yellow eyes; male has purple sheen to head,
female is slightly grayer below; juveniles indistinguishable from
juvenile Boat-tailed. Common Grackle shorter-tailed, has yellow eyes,
and purple or greenish gloss on back and wings.

Range
Resident in coastal areas from New York to central Texas and
throughout Florida peninsula. Occasionally along rivers inland.

Bronzed Cowbird
Molothrus aeneus

Robin-sized
Southern Texas: Open Country;
Residential Areas

Field Marks
8¾". In southern Texas, these cowbirds are found in riparian
woodlands, brushy areas, fields, feedlots, and towns. Red eye is
diagnostic. Male performs a bizarre helicopter flight and prancing
display.
Adult males (1) black with metallic sheen and ruff behind head; bill
deep and relatively long; crown rather flat; eyes deep red. Adult
females (2) similar, but have back flat, without ruff. Juveniles gray-
brown, with dark eye.
Call a low *chuck*; also squeaky notes like a rusty hinge.

Similar Species
Deep red eye and ruff distinctive. Brown-headed Cowbird smaller,
rounder, with smaller bill. Rusty and Brewer's blackbirds slimmer,
with smaller, slender bills, light eyes. Grackles much larger, with
light eyes.

Range
Breeds from s. Arizona to s. Texas, south to Central America.

Brown-headed Cowbird
Molothrus ater

Sparrow-sized
Open Country and Grasslands;
Groves

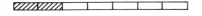

Field Marks
7½". Relatively small black bird with rounded proportions and a
short, conical bill. Bill shape and short, squared tail distinguish it from
blackbirds. Often flocks with blackbirds and starlings in residential
and agricultural areas. In mixed-species flocks on the ground, note
habit of sticking its tail straight up when feeding.
Males (1) glossy, iridescent black with rich brown head. Females (2)
uniformly brownish-gray. Juveniles pale gray, streaked with dark
brownish-gray. Tail short and squared at tip in all plumages.
Male's song consists of explosive squeaking and bubbling notes. Flight
calls high-pitched and squeaky; also a guttural rattle.

Similar Species
Brown head diagnostic. See Bronzed Cowbird.

Range
Breeds from ne. British Columbia east to Newfoundland and south to
Texas. Winters across s. United States.

Fish Crow
Corvus ossifragus

Crow-sized
Beaches and Coastal Dunes; Salt
Marshes; Freshwater Marshes;
Inshore Waters; Mudflats; Rivers

Field Marks
15½". A common crow in maritime areas along the Atlantic and Gulf coasts; also moves upstream on large rivers. Distinguished by its nasal call, but otherwise closely resembles American Crow. Nests in colonies. Feeds on carrion, aquatic animals, and insects; also at garbage dumps.
Adults (1) entirely glossy black, with subtle slate-violet sheen on the wing. The tail is slightly rounded at the tip. Immatures similar.
Call a high, nasal *bah*, or doubled *uh-uh*, more abrupt than American Crow's call.

Similar Species
American Crow best distinguished by call, but is also somewhat larger; flies on less rapid wingbeats. The Mexican Crow, *Corvus imparatus*, visitor to southern Texas, is smaller (14"); has low, froglike call and slightly more lustrous, dark violet-blue sheen on upperparts.

Range
Breeds along Mississippi River and northern Gulf Coast, and along Atlantic Coast from s. New England to Florida. Some northern birds move south in winter.

American Crow
Corvus brachyrhynchos

Crow-sized
All forest, open, urban, and
residential habitats; Rocky
Shores; Beaches

Field Marks
17½". A familiar species found just about everywhere, but in very
large numbers in areas associated with man's activities, such as
farmlands, towns, and rural roadsides. Highly gregarious; can occur
in very large flocks, especially at nightly roosting areas in fall and
winter.
Adults (1) entirely black, with rather stout bills and legs; tail squared
and slightly fan-shaped. Immatures resemble adults.
Call a familiar, slightly descending *caw*.

Similar Species
Fish Crow has nasal, single *bah*, *uh-uh*, or *au-ought* note; young
American Crow has similar quality note, but more of a *caah*. Larger
Common Raven has much larger bill, wedge-shaped tail. Mexican
Crow smaller, glossy purplish-black; call higher-pitched.

Range
Breeds from north-central British Columbia east to Newfoundland,
south to California, e. Texas, and Florida. Winters throughout most of
United States.

Chihuahuan Raven
Corvus cryptoleucus

Crow-sized
Dry Open Country

Field Marks
19½″. A small raven found in open, arid grasslands in Southwest and parts of Great Plains; often occurs with yucca, mesquite, and other brushy growth. Difficult to distinguish from larger Common Raven, which usually occurs at higher elevations. Chihuahuan Ravens often form large flocks; soar on rising air currents, or thermals.
Adults (1) black overall, with rather long, wedge-shaped tail and deep, stout bill. Feathers on neck and underparts are white at base; white diagnostic, but usually not visible. Immatures resemble adults. Call a hoarse, flat *kraaak*.

Similar Species
American Crow has smaller bill, squared and slightly fan-shaped tail; *caw* call. Common Raven larger, with deeper, slower wingbeats; has much deeper *krock* call; also a much higher call than Chihuahuan: a rough, screaming *aack*.

Range
Se. Arizona east to w. Kansas and south to Mexico. Northern birds move south in winter.

Common Raven
Corvus corax

Crow-sized
Boreal Forest; Coastal Cliffs

Field Marks
24″. A large, all-black bird, similar to American Crow, but larger and heavier, with a much larger bill, longer wings, and long, wedge-shaped tail. Once seriously declining in eastern United States, but now Common Ravens are increasing rapidly in the Adirondacks and Appalachians. In the East, occurs in boreal forests and along northern seacoasts.

Adults (1) large; glossy black overall; very deep bill has distinctively curved upper surface; shaggy feathering on neck; wings broad, long, and rounded toward tip; tail long, full, and wedge-shaped. Immatures similar to adults.

Calls are a low, guttural *krock* and a high-pitched, screaming *aack*; also a rolling, throaty rattle. Much variation.

Similar Species
See Chihuahuan Raven, American Crow.

Range
Resident from n. Alaska east to Newfoundland and south through the West, Great Lakes region, and the Northeast; also in Appalachians.

Smooth-billed Ani
Crotophaga ani

Pigeon-sized
Southern Florida: Open Country;
Thickets

Field Marks
14½″. An all-black cuckoo with a long, loose tail and a large, rounded
bill that is laterally compressed like a puffin's bill. When flushed,
rapidly flaps its short, rounded wings, then briefly glides and gives
another series of quick flaps. Fairly common resident in farmland,
scrubland, and cattle pastures; eats mainly insects. Distinctive bill has
a steeply curved ridge on the upper mandible.
Adults (1) dull to glossy black; feathers of upper back and breast with
pale V-shaped fringes; nape feathers tipped with bronze.
Call a whining, metallic, rising whistle: *wee-ick*.

Similar Species
Groove-billed Ani, *Crotophaga sulcirostris* (2), of Texas and
Louisiana, similar but lacks arched ridge on culmen, has grooves on
bill, and gives a liquid, descending call. Grackles have more slender
bills and steady, flapping flight.

Range
Resident in s. Florida; occasionally wanders to other states along Gulf
and Atlantic coasts.

Black-billed Magpie
Pica pica

Pigeon-sized
Open Country; Groves

1

Field Marks
19″. An unmistakable, striking, black-and-white-plumaged bird with
long, tapered tail. Inhabits open country and riparian woodlands.
Common, conspicuous, and gregarious; frequently seen along
highways. Takes a variety of insects, but also feeds on carrion,
especially at road kills.
Adults (1) have iridescent black upperparts, head, neck, and lower
belly; white on rest of underparts gives pied appearance; white
shoulders and wing patches flash in flight; iridescent black tail very
long and tapered. Immatures similar to adults.
Call a rising, inquisitive-sounding *maag*; also a rapid, harsh *queek
queek queek*.

Range
Resident from south-central Alaska east to Manitoba and the Great
Plains, south to n. California and n. New Mexico.

Brown Jay
Cyanocorax morio

Pigeon-sized
Southern Texas: Streamside
Forest

1

Field Marks
16½″. A large, long-tailed, but crowlike jay, uncommon and local within the United States. Inhabits streamside woodlands; often travels in small, noisy bands. Has a hidden throat pouch that it uses to make a popping noise in combination with calls. Flight appears slow and labored. Large size, heavy bill, and vocalizations distinctive. Adults (1) mainly dark, sooty brown, with whitish belly and undertail coverts; tail long, wide; bill black; eyes dark. Young birds have a yellow bill and fleshy ring around eye; bill darkens with age and may be mottled in some individuals.
Call a loud, explosive *pow* or *kee-ow*.

Similar Species
Magpies are black, with white on underparts and in wings; have pointed tail; range does not overlap. Thrashers much smaller; none has plain white on belly.

Range
Resident but local in lower Rio Grande valley of s. Texas. Also in Mexico and Central America.

Gray Jay
Perisoreus canadensis

Robin-sized
Boreal Forest

Field Marks
11½". Unmistakable; looks like giant chickadee. A fluffy, quiet bird of
boreal coniferous forests. Generally quite tame; not noisy, raucous, or
conspicuous like most jays. Small groups fly silently from tree to tree
with direct, level flight, alternating rapid wingbeats with glides.
Rarely seen singly.
Adults (1, 2) have dark gray upperparts and crown; whitish face,
turning to light gray on underparts. Most eastern birds have dark
nape; one race (2) is paler on nape. Juveniles uniformly dark gray with
whitish whisker mark.
Song very different from other jays; consists of one to several loud,
mellow, clear whistled notes. Also a quiet, harsh *shack* note.
Generally quiet.

Range
Resident from Alaska east to Newfoundland, south to n. California
and n. New Mexico; in East, to northern Great Lakes region and
n. New England.

Scrub Jay
Aphelocoma coerulescens

Robin-sized
Florida: Forest habitats; Thickets

Field Marks
11½". A crestless blue-and-gray jay found in scrub-oak barrens; in the East, occurs only in peninsular Florida. These social birds are noisy and conspicuous; they characteristically jerk the body up and down with each call note. Bounding flight involves bouts of rapid flapping, alternating with sweeping, gliding arcs.
Adults (1) have blue head, wings, rump, and tail; throat and forehead whitish with grayish streaking; eyeline white; ear patch dusky; back and underparts grayish, with vague blue necklace on breast.
Calls varied. Common ones include a raucous, slightly metallic, sharply inflected *iennk* and a rapid series of harsh *quick-quick-quick* notes.

Similar Species
Blue Jay has crest, black necklace, black bars and white spots in wings and tail.

Range
Resident from sw. Washington east to Wyoming and Colorado, south to s. California and central Texas. An isolated population established in central Florida.

Blue Jay
Cyanocitta cristata

Robin-sized
Forest habitats; Residential Areas
and Parks

Field Marks
11½". A handsome, crested jay that for most of the year is found in loose, noisy groups traveling through woodlands and parks. Highly vocal in fall, when its harsh, nasal *jay* is a familiar sound. Becomes shy and somewhat secretive during breeding season. Eats a variety of insects, fruits, and seeds; haphazardly stores acorns and nuts for winter use.

Adults (1) bright blue above with white spots and black bars on wings and tail; back has a slight purplish-gray cast; whitish on face and below, with black necklace around upper breast and back of head; breast and sides washed with gray.

Calls extremely variable, from harsh to musical; include a harsh, nasal *jay*, sometimes slurred; and a liquid, whistled *whee-oodle* and *oolink*.

Range
Mainly resident from central Alberta east to Newfoundland and south to e. Colorado, e. New Mexico, se. Texas, and Florida; rarely in Northwest. Northern birds may withdraw in winter.

Green Jay
Cyanocorax yncas

Robin-sized
Southern Texas: Streamside
Forest

Field Marks
10½". A colorful jay found in much of the tropics, but only locally within the United States in extreme southern Texas. Favors the cover of woodlands; quite adept at staying concealed. Typically found in loose flocks.

Adults (1, 2) unmistakable. Brilliant green above with a shiny, powder-blue head and face; black eyebrow connects with a broad black patch from the ear to the chin and throat; spot above the eye, malar area, and rear of eye blue; underparts pale yellow-green; tip of green tail tinged blue; outer feathers bright yellow.

Call a rapid *cheh-cheh-cheh* and slower *cleep, cleep, cleep*; or a wide variety of notes repeated two or three times. Also gives low, throaty rattles.

Range
Resident in lower Rio Grande valley of s. Texas. Also in tropical America.

Red-whiskered Bulbul
Pycnonotus jocosus

Sparrow-sized
Southern Florida: Residential
Areas and Parks

Field Marks
7". A crested bird with a bright scarlet face patch. Locally established as a small breeding population following escape from captivity in Miami, Florida; found in parks and gardens in this same area. Feeds mainly on small fruits.

Adults (1) dark brown above with tall, blackish crest; have large red spot behind eye and red undertail coverts; dark whisker connects with the dark neck; underparts otherwise white; tail tipped with white; bill thin, black. Immatures lack red spot behind eye; have duller upperparts and undertail coverts.

Song a loud, clear, rising and falling whistle: *queek-kee* or *queep-kwill-ya*, somewhat like American Robin's song. Also gives chattering notes.

Similar Species
Female Northern Cardinal has red wings and crest; bill thick and conical.

Range
Introduced and resident in s. Florida.

Bohemian Waxwing
Bombycilla garrulus

Robin-sized
Boreal Forest; Eastern Deciduous
Forest

Field Marks
8¼″. A lovely, large, dark edition of the familiar Cedar Waxwing; in the East, encountered only in winter. Gregarious and erratic in occurrence; frequents orchards and berry-producing trees planted in rural areas.
Adults (1, 2) basically gray, darker on upperparts, with chestnut on undertail coverts and around black face patch and throat; white, yellow, and red patches in wing; tip of tail yellow; long, slender, elegant crest. Juvenile similar but paler overall, with streaked underparts.
Call a soft, prolonged, high-pitched buzzy note; lower and more buzzy than Cedar Waxwing's.

Similar Species
Cedar Waxwing smaller, warm brown, with a yellow belly and white, not chestnut, undertail coverts.

Range
Breeds from central Alaska east to n. Manitoba and south to Washington, Idaho, and Montana. Winters from Washington east to n. Maine; occasionally to California and n. Texas.

376

Cedar Waxwing
Bombycilla cedrorum

Sparrow-sized
Forest habitats; Open Country;
Residential Areas and Parks

Field Marks
7¼". A sleek, elegant, soft-plumaged bird common throughout the
United States in winter. Highly gregarious; often in flocks (together
with American Robins) around residential areas where berry-
producing shrubs and trees abound. Waxwings constantly give soft,
very high-pitched notes, and flock members busily fly back and forth
from high perches to nearby bushes or trees with berries. Also hawks
for flying insects. Flies in tight flocks on rapid wingbeats with
alternating short, arching glides. Open wing relatively triangular, and
tail short and squared.
Adults (1, 2) warm brown turning to yellow on belly and white on
undertail coverts; rump, wings, and tail gray, with waxy red tips on
secondaries, and yellow band at tip of tail; black mask bordered
narrowly by white. Juveniles similar but duller, streaked below.
Calls are soft, high-pitched, prolonged single notes, often clear but
sometimes with lisping quality.

Similar Species
Bohemian Waxwing larger, grayer; has chestnut, not white or yellow,
on lower underparts; also has white and yellow in wing.

Range
Breeds from se. Alaska east to Newfoundland, across n. United
States from n. California to n. Georgia. Winters from s. British
Columbia east to Maine and south throughout United States.

Loggerhead Shrike
Lanius ludovicianus

Robin-sized
Thickets; Open Country

Field Marks
9″. A gray, black, and white predator with a black mask and dark, hooded bill. Inhabits open country; becoming quite rare in the Northeast. Perches conspicuously on fences, wires, and treetops to search for insects, birds, and small rodents. Flies low and direct with rapid wingbeats; white patches flash in black wings.
Adults (1–3) primarily gray, darker on upperparts, with black bill and broad black mask extending from bill to behind eye; wings black with white patches; tail black, medium-length to long, and slightly rounded, with white outer tail feathers. Juvenile (late summer) similar, but barred with gray on underparts, brownish in wings. Song a series of soft, somewhat musical warbles interspersed with harsh, buzzy squeaks. Calls include a mechanical *zee-eut* and a harsh *sheek sheek*.

Similar Species
See Northern Shrike. Northern Mockingbird has thin bill, lacks mask.

Range
Breeds from central Alberta east to se. Ontario and south to Mexico and Florida; absent from most of Northeast. Winters across southern half of United States.

Northern Shrike
Lanius excubitor

Robin-sized
Boreal Forest; Open Country

Field Marks
10″. The northern equivalent of the Loggerhead Shrike, and the only shrike usually seen in Northeast in winter. Perches conspicuously in open country or open woods; often in trees.

Adults (1, 3) have light-medium gray upperparts with whitish rump; underparts white, lightly barred with grayish; bill rather long, deeply hooked, often has pale base to lower mandible; dark mask, primarily behind eye; wings black with white patches; tail black, relatively long with white in outer features. Immature (2) in first winter similar but browner, with stronger barring on underparts.

Song a variable series of musical warbles interspersed with harsh, metallic phrases. Call a harsh *jaaeg* and *shek-shek*.

Similar Species
Loggerhead Shrike smaller; shorter bill lacks pale base; less frosty gray overall; has relatively larger face mask extending from bill to behind eye; underparts lack barring; rump gray. Young birds have less extensive brownish tones than young Northern Shrikes, particularly juvenile Northern. See also Northern Mockingbird, which has thin bill, lacks mask.

Range
Breeds from Alaska east to Labrador. Winters from s. Alaska across n. United States.

Snow Bunting
Plectrophenax nivalis

Sparrow-sized
Open Country; Beaches and
Coastal Dunes

Field Marks
6¾". A common, hardy breeder of far northern tundra. Plumage
changes twice a year: once by molt, once by wear. Breeding male
unmistakable. Black-and-white wings conspicuous in flight in both
sexes year-round. Gregarious; often flocks with Horned Larks and
longspurs in winter. Feeds on ground in open country.
Breeding male (2) white with dark back, wing tips, and central tail
feathers. Breeding female (1) duller white with finely streaked,
grayish crown; dark streaked back; tail like male's; wing with less
white. Freshly molted male (3) with black upperparts broadly edged
white or buff; crown and ear patch rusty brown; mostly white below.
Winter female resembles male, but darker above, duller below. In
both sexes, feather wear produces breeding plumage by summer.
Immatures resemble adults.
Calls include a sweet *tew* or *tew-eee*; a short trill; a raspy *wreent*.
Warbling song heard only on nesting grounds.

Similar Species
Female and winter male Lark Bunting streaked below; have
heavy bills.

Range
Breeds in n. Alaska and n. Canada. Winters south to Oregon and
the Carolinas.

380

Lark Bunting
Calamospiza melanocorys

Sparrow-sized
Brushy Open Country and
Grasslands

Field Marks

7″. A bird of prairies, fields, and brushy areas; gregarious in winter.
Black-and-white breeding male unmistakable; females, immatures,
and winter males streaky and sparrowlike. Bill heavier than that of
similar-looking streak-breasted sparrows.
Breeding male (1) all-black with white patches on inner wing
conspicuous perched and in flight. Breeding and winter female (2) and
immature brown above with dark streaks, whitish below with dark
streaks; show pale buff or whitish wing patches, brown cheek patch
with light outline. Winter male (3) like female but streaked darker
below, with black chin.
Song, often given in flight, a mixture of varied, rich notes, most
repeated several times in mockingbirdlike pattern; also a two-noted
whistle call.

Similar Species

Winter male, female, and immature told from streak-breasted
sparrows by combination of heavy bill, pale wing patch, and brown
cheek patch.

Range

Breeds from British Columbia and Manitoba south through the Great
Plains to n. New Mexico and n. Texas. Winters from s. Arizona to
central Texas.

381

Rose-breasted Grosbeak
Pheucticus ludovicianus

Robin-sized
Eastern Deciduous Forest;
Thickets; Residential Areas

Field Marks

8″. A heavy-billed, black-and-white bird with a musical song. Found in eastern forests as well as in thickets and suburbs with suitable cover. Males help to incubate eggs during nesting season, singing from the nest. All plumages have stout, pale, conical bill.

Breeding males (1) have black head and upperparts, white belly, white wing patches and bars; breast with a bright splash of red; in flight, shows red wing linings and white rump. Females (2) dark brown above, white below, with dark brown spots and streaks; head dark brown with distinctive white eyebrow; yellow or pinkish-yellow wing linings visible in flight. Winter males similar to females, but usually with some trace of red on breast; red wing linings and black-and-white flight feathers visible in flight. Immatures similar to females, but show considerable variation.

Song rich, clear, like a robin's but faster. Also *ink* and *eek* notes.

Similar Species

Female Black-headed Grosbeak, *Pheucticus melanocephalus*, rare visitor from West, has buff or orange wash on breast; paler yellow wing linings. See female Purple Finch.

Range

Breeds from southern Mackenzie east to Nova Scotia and south to ne. Oklahoma, n. New Jersey, and in mountains to n. Georgia. Winters from Mexico to South America; occasionally in Southwest.

White-winged Crossbill
Loxia leucoptera

Sparrow-sized
Boreal Forest; Pine Barrens;
Other forest habitats

Field Marks
6½″. A sparrow-sized finch, widespread in coniferous and mixed
coniferous-deciduous forest. Two white wing bars conspicuous in all
plumages. Like many northern finches, wanders irregularly,
depending on availability of cones; irrupts far south of usual range in
some winters. Distinctive crossed mandibles are used to pry open
cones; obvious only at close range. Often vocal in flight. May nest as
early as January.
Adult males (1) bright pinkish-red with dark tail; dark wings have two
white wing bars. Adult females (2) have fine dark streaks above and
below, olive-gray with dark tail; rump and underparts dull yellow;
dark wings with two white wing bars. Immatures similar to adult
female but duller, conspicuously streaked; wing bars may be
narrower.
Flight call a series of *chit* notes; song a series of monotone trills and
buzzes, each on abruptly different pitch.

Similar Species
Red Crossbill darker red, lacks white wing bars. Male Pine Grosbeak
larger; lacks crossed mandibles; white in wings less conspicuous.

Range
N. Alaska east to Newfoundland and south to British Columbia and
central Maine. Winters south across northern half of United States.

Pine Grosbeak
Pinicola enucleator

Robin-sized
Boreal Forest; Other forest
habitats; Brushy Open Country

Field Marks
9″. A large, plump finch of coniferous forest and edges; in migration and winter also in deciduous habitats and second growth. Two white wing bars conspicuous in all plumages. Tail long; bill small, conical; flight undulating, like other finches. Feeds in trees and on ground; consumes seeds, buds, fruit. May irrupt far south of usual range in winters when food is scarce.

Adult males (1) pinkish-red with long, dark tail; dark wings with two white wing bars; gray flanks and belly. Females (2) gray with yellow-olive (some rust-tinged) head and rump; dark tail; dark wings with two white wing bars. Immature males similar to females but with dull rusty head and rump.

Flight call a distinctive, whistled *pew-pew-pew*. Song variable, reminiscent of Purple Finch or House Finch but not as burry, and less varied in pitch.

Similar Species
White-winged Crossbill smaller; has crossed mandibles.

Range
Alaska east to Newfoundland and south in mountains to central California. Winters south to North Dakota, New York, and n. New England.

Red Crossbill
Loxia curvirostra

Sparrow-sized
Boreal Forest; Pine Barrens;
Other forest habitats

Field Marks
6¼". A sparrow-sized finch of coniferous and mixed coniferous-deciduous forests of the North and West. Like many northern finches, wanders irregularly, depending on availability of pinecones, its principal food; irrupts far south of usual range during some winters. Distinctive crossed mandibles, used to pry open cones, obvious only at close range. Often vocal in flight. Nests as early as January. Plumage variable. Typical adult males (1) dull orange-red with dark wings and tail. Typical adult females (3) grayish-olive or yellowish-olive with dull yellow rump and underparts; dark wings and tail. Immatures (2) resemble adults but duller and streaked.
Flight call usually two or three *jip* or *kip* notes. Song of whistled, trilled, warbled notes and phrases, often doubled or tripled.

Similar Species
Pine Siskin similar to streaked immature, but smaller; lacks crossed mandibles; has yellow in wings and tail. White-winged Crossbill shows two white wing bars in all plumages.

Range
Breeds from se. Alaska east to Newfoundland and south to northern Great Lakes region, and northern New York and New England. Winters south to central United States.

Scarlet Tanager
Piranga olivacea

Robin-sized
Eastern Deciduous Forest; Other
forest habitats; Residential Areas
and Parks

Field Marks
7¾". A medium-sized, stout-billed songbird of deciduous and mixed
pine-oak forests; frequently found in shade trees and well-planted
suburban areas. Usually forages sluggishly and inconspicuously in
tree canopy, where loud song is best indication of its presence.
Adult males (2) brilliant red with black wings and tail. Females (1)
olive-green above and yellowish below, with dull olive to dusky black
wings and grayish bills. Winter males and immatures resemble
females; molting males have bodies blotched with red and yellow.
Immatures may show faint wing bars.
Song a hurried, burry, robinlike series of four to five phrases; also a
loud, twanging *chipp-burr*.

Similar Species
Female Summer Tanager resembles female Scarlet Tanager, but
usually more orange-yellow with less contrasting wings and a larger,
paler bill.

Range
Breeds from s. Manitoba east to Maine and south to e. Oklahoma and
w. North Carolina. Winters in South America.

Summer Tanager
Piranga rubra

Robin-sized
Eastern Deciduous Forest

Field Marks
7½″. Widespread across southern United States, favoring mixed
deciduous woodland and pines in the East, cottonwoods and willows in
the West. Feeds on insects high in trees, where bright males remain
concealed but are often heard.

Adult males (2) wholly red, wings slightly darker. Females (1) olive to
grayish-olive above (including wings and tail), orange-yellow below
(darkest in East); some eastern females patchily reddish. Young males
resemble females until first spring, when become patchy to mostly
red; all red by second fall. All birds have pale bill.

Distinctive call a three-syllabled, rapid *pitty-tuck* rattle. Song
robinlike, slightly less burry than Scarlet Tanager's.

Similar Species
Female Scarlet Tanager more greenish above, yellow-green (not
orange) below.

Range
Breeds from se. California to central Oklahoma; s. Nebraska and s.
New Jersey south to Gulf Coast; occasionally in Northeast. Winters
from Mexico to South America.

Pyrrhuloxia
Cardinalis sinuatus

Robin-sized
Southern Texas: Streamside
Forest; Brushy Open Country

Field Marks
8¾". A cardinal-like bird of mesquite, weedy fields, and streamside thickets; often in small flocks in winter. Downcurved, parrotlike bill distinctive in all plumages.
Adult males (2, 3) mostly gray with red-tipped crest; tail, primaries, and thighs red; red patch around eyes and bill extending down throat and center of breast; bill yellow. Adult females (1) grayish above; buff-gray below; may show some pink to dull red on crest, tail, primaries, thighs, and around eyes; bill dull yellow. Immature similar to female but with dark bill.
Song a series of loud, clear whistles; similar to Northern Cardinal's but usually thinner, shorter.

Similar Species
Female and immature Northern Cardinal usually not as gray; bill conical, not downcurved.

Range
Resident from south-central Arizona east to s. Texas; rarely in se. California.

Northern Cardinal
Cardinalis cardinalis

Robin-sized
Eastern Deciduous Forest;
Thickets; Residential Areas and
Parks

Field Marks
8¾″. One of most familiar North American songbirds, widespread and abundant in woodlands, forest edges, brushy areas, and residential areas. Adult male unmistakable. Both sexes sing almost year-round. Nonmigratory. Feeds on ground or in shrubs and trees on insects, fruit, seeds, grain; also visits feeders.
Adult males (1) bright red with conspicuous crest; bill large, conical, red; face and throat black. Adult females (2) grayish-olive above, buff below; variably red on wings, tail, and crest; face dark; bill pink. Immature similar to female but browner (immature female often lacks red); bill dark.
Song a variable series of loud, clear whistles; usually includes a rapid *cheer-cheer-cheer* and *whoit-whoit-whoit.*

Similar Species
Female and immature Pyrrhuloxia usually grayer than female and immature cardinal; downcurved, parrotlike bill distinctive.

Range
Resident throughout e. United States; also from s. Arizona east to sw. Texas.

Painted Bunting
Passerina ciris

Sparrow-sized
Thickets; Brushy Open Country;
Residential Areas and Parks

Field Marks
5½". Unmistakable male is one of brightest North American birds but
often stays concealed; greenish female easily told from closely related
Indigo, Lazuli, and Varied buntings. Nests in thickets, gardens, and
open areas with scattered trees and shrubs; also found in light woods,
open and shrubby habitats in migration and winter (mostly south of
United States).
Adult males (1, 3), brightest in spring and summer, have blue head,
yellow-green back; wings and tail blackish; rump, eye-ring, and
underparts red. Females (2) dull green, paler greenish below with
buff or yellowish cast. Immature males resemble females (sometimes
with hint of adult pattern) for first year.
Song a burry warble, more slurred than songs of other buntings;
reminiscent of Warbling Vireo but higher-pitched, softer, with notes
jumping less in pitch.

Similar Species
Female and young Indigo, Lazuli, and Varied buntings brownish.

Range
Breeds from se. New Mexico east to s. Missouri and the Gulf Coast;
also along southern Atlantic Coast. May wander farther north.
Winters from se. Texas, s. Louisiana, and central Florida to
Central America.

Lazuli Bunting
Passerina amoena

Sparrow-sized
Thickets; Brushy Open Country

Field Marks
5½". Replaces closely related Indigo Bunting in most of West. Nests in brushy areas and open woodland, often near water; shrubby and open areas in migration and winter.

Adult males (1) have blue upperparts (back mottled with gray) and throat; tail blackish; wings blackish with two white wing bars; breast rusty; belly white. First-year and some winter adult males duller, clouded with brown above. Adult females (2) and immatures brown, paler below; wing bars buff; in adult females wings, rump, and tail have more or less bluish cast.

Song a lively jumble of warbled (single, doubled, or tripled) notes or short phrases; similar to Indigo Bunting's but usually faster.

Similar Species
Male Eastern Bluebird lacks wing bars, has slender bill. Indigo Bunting has breast streaks. Varied Bunting lacks wing bars. Female Painted Bunting greenish. Female Blue Grosbeak larger with more prominent wing bars; bill much heavier.

Range
Breeds from British Columbia east to North Dakota and south to California and w. Oklahoma; occasionally farther east. Winters from s. California and s. Arizona south to Mexico.

Varied Bunting

Passerina versicolor

Sparrow-sized
Southern Texas: Brushy Open
Country

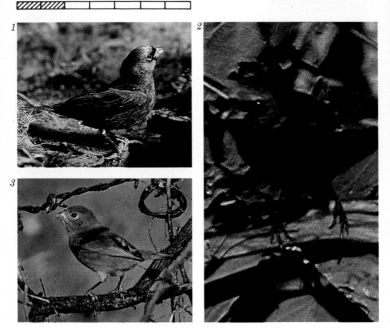

Field Marks

5½". Favors dry scrub and thickets in breeding range, which barely
extends into southwestern United States. Male often looks blackish;
subtle but handsome colors obvious only in good light.

Breeding adult males (1, 2) mostly dark with reddish-purple cast on
back and breast; rump and variable forehead and collar blue; patch on
back of head rust-red. Females (3) all-brown with darker wings and
tail; lack obvious wing bars. Winter adult male has breeding pattern
partially obscured by brown edging, which disappears by spring.
Immature males resemble females until second winter.

Warbled song like Lazuli or Indigo bunting's, but often shorter.

Similar Species

Female resembles female Indigo and Lazuli buntings: Indigo has
breast streaks; Lazuli has prominent wing bars. Female Painted
Bunting greenish. Female Blue Grosbeak larger; wing bars more
prominent; bill much heavier.

Range

Breeds locally in se. Arizona, New Mexico, and s. Texas. Winters
south of United States.

Eastern Bluebird
Sialia sialis

Sparrow-sized
Open Country

Field Marks
7″. Unmistakable; only small, bright blue bird in most of the East.
Inhabits open country with scattered trees, fence posts, and telephone
poles. Nests in cavities, and will make use of nest boxes where
competition with European Starlings and wrens is not fierce. Often
seen perched on a tree in hunched posture.
Adult males (1) have chestnut throat and breast, white belly, and
bright blue head, nape, back, and wings. Adult females (2) similar but
paler overall, with brownish back. Immatures like females but have
darker head and brown streaks and spots on breast.
Song a very musical, liquid *chur-lee* or *tur-ee*, repeated in warbled
phrases.

Similar Species
In Great Plains, overlaps with more western Mountain Bluebird,
Sialia currucoides; male entirely sky-blue, with no rust or chestnut in
plumage; female paler, grayer, with no strong reddish or chestnut
colors in plumage. Western Bluebird, *Sialia mexicana*, has gray
throat, gray belly; male Western has chestnut patch on back. Lazuli
Bunting is much smaller, with conical bill and two white wing bars.

Range
Breeds throughout eastern North America west to the Great Plains;
also in mountains of se. Arizona and w. New Mexico. Winters
throughout se. United States.

393

Blue Grosbeak
Guiraca caerulea

Sparrow-sized
Thickets; Brushy Open Country

Field Marks
6¾". A small grosbeak of old fields, streamside thickets, and brushy open country. Has distinctive, large, conical bill, typical of grosbeaks. Usually forages on ground, primarily on insects, grain, seeds, fruit. Flicks tail in a figure eight.

Adult males (1) rich blue (may look black in poor light) with two chestnut wing bars. Adult females (2) brown with two chestnut or buff wing bars; rump pale blue. Immatures similar to female. First-spring male patchy blue and brown.

Song a series of short, sweet warbles; similar to Purple Finch's song, but slower.

Similar Species
Indigo Bunting smaller; lacks wing bars; bill smaller. Female and immature Brown-headed Cowbird lack wing bars; do not flick tails.

Range
Breeds from central California east to s. New Jersey, north to South Dakota and south to Texas and n. Florida. Winters in Mexico and Central America.

Indigo Bunting
Passerina cyanea

Sparrow-sized
Thickets; Brushy Open Country

Field Marks
5½". Replaces closely related Lazuli Bunting in East. A common
nester in shrubby woodland edges, regenerating pastures, power-line
cuts; also light woods, open and shrubby habitats during migration
and winter.
Breeding adult males (1) blue (appear blackish in poor light); wings
and tail mixed with black. First-spring males mottled with brown.
Females (2) and first-winter males brown (paler below), with faint
buff wing bars and faint dark breast streaks; wings and tail with or
without bluish cast.
Song a lively jumble of warbled (often repeated) notes or short
phrases; similar to Lazuli Bunting's but usually slower.

Similar Species
Blue Grosbeak and Mountain Bluebird resemble all-blue breeding
male but larger. Female Lazuli and Varied buntings lack streaks;
Varied lacks wing bars. Female Painted Bunting greenish. Female
Blue Grosbeak larger; unstreaked; bill much heavier. Mountain
Bluebird larger; bill finer, longer.

Range
Breeds from se. Manitoba east to central Maine and south to central
Texas and n. Florida. Winters in s. Florida and occasionally along Gulf
Coast, primarily in Central America.

Evening Grosbeak
Coccothraustes vespertinus

Robin-sized
Boreal Forest; Eastern Deciduous
Forest; Residential Areas and
Parks

Field Marks
8″. A large, plump finch of coniferous and mixed coniferous-deciduous forests and second growth; in migration and winter also around residential areas. Unmistakable, with very large conical bill, short black tail, black wings with large white patches. Gregarious. Flocks wander irregularly in winter, depending on availability of food. Feeds heavily on buds and seeds of conifers and maples.

Adult males (1, bottom) have brown head with yellow forehead and eyebrow; yellow back, rump, and belly; wings black with large white patches; tail black; bill yellow. Adult females (1, top) grayish with yellow wash; wings and tail black with white markings; bill greenish-yellow. Both sexes show duller bill color in winter. Immatures resemble adults by first winter.

Song a short warble; more commonly heard call note a shrill *pe-teer*.

Range
Breeds from north-central British Columbia east to Nova Scotia, south to Great Lakes region and n. New England, in mountains to central California. Winters throughout most of United States.

American Goldfinch
Carduelis tristis

Sparrow-sized
Brushy Open Country; Groves;
Residential Areas and Parks

Field Marks
5″. A small, common finch, widely distributed in weedy fields, open
woodlands, forest edges, and second growth. Often around houses;
common at feeders. Typically in flocks; often quite vocal. As in many
finches, flight bounding. Eats mostly seeds, especially thistle.
Breeding males (1) bright yellow with black cap; black tail with white
undertail and upper-tail coverts; black wings with two narrow white
wing bars. Breeding females (2) mostly yellow (brightest on throat
and breast); wings and tail similar to male's but duller. Winter male
(3) and female brownish or grayish above (male with stronger yellow
wash on head); wings black with two buff or dull white wing bars;
underparts buff or pale grayish. Immature similar to breeding female
but cinnamon above and on wing bars.
Flight call a distinctive *per-chick-o-ree*. High-pitched, jumbled, and
twittering song best distinguished by up-slurred *su-wee* or *swee* notes.

Similar Species
Lesser Goldfinch green or black on back; shows white wing patch at
base of primaries; undertail coverts usually yellow. Immature Indigo
Bunting streaked below.

Range
Breeds from British Columbia east to Nova Scotia and south to s.
California, southern Great Plains, and n. Georgia. Winters south to
Mexico and Florida, rarely in interior of United States.

Lesser Goldfinch
Carduelis psaltria

Very Small
Texas: Brushy Open Country

Field Marks
4½". A small finch of weedy fields, woodland edges, and second
growth; often seen near houses. Male shows two color forms,
green-backed and black-backed. Typically in small flocks. Eats mostly
seeds.

Adult males (2) have green or black back (black plumage less common;
usually seen on birds in eastern part of range); cap black; wings black
with white edging and white patch (most visible in flight) at base of
primaries; tail black with inconspicuous white at base; underparts
bright yellow. Females (1, 3) greenish above, with white in wings and
tail visible in flight; dull yellow below. Immatures similar to females
but greener below.

Song a high-pitched jumble of twittering notes, not as sweet as
American Goldfinch's; calls a falling *teee* and a rising or falling *tee-eee*.

Similar Species
American Goldfinch not greenish; lacks white patch at base of
primaries; undertail coverts white (not yellow).

Range
Resident from sw. Washington east to Colorado and central Texas,
south to n. South America. Occasionally wanders farther east.

Olive Sparrow
Arremonops rufivirgatus

Sparrow-sized
Southern Texas: Streamside
Forest; Thickets

Field Marks
6¼". A plain-looking, somewhat secretive Central American sparrow;
in our range, occurs in thickets and streamside vegetation in southern
Texas. Feeds on small insects and larvae, foraging on the ground;
scratches in dry leaves like towhee; also flies quickly from one
shrubby bush to another. Sings from exposed perches.
Adults (1) plain olive above and below, somewhat paler on underparts,
with two reddish-brown stripes on crown, and narrow line through
eye. Immatures similar, with more extensive brown on crown.
Song a succession of dry *chip* notes. Also gives a harsh *tick*.

Similar Species
Green-tailed Towhee, *Pipilo chlorurus*, winter visitor from western
mountains, larger, with grayish face and breast, and chestnut crown;
immature streaked above and below; white throat with black outline
in both plumages.

Range
Resident in s. Texas; also in Mexico and Central America.

Rufous-sided Towhee
Pipilo erythrophthalmus

Robin-sized
Forest habitats; Thickets

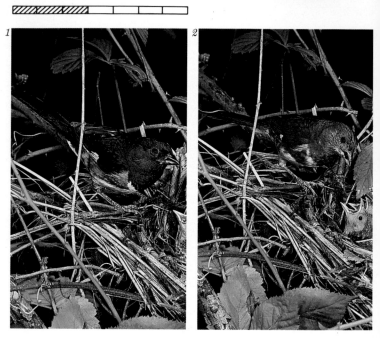

Field Marks
8½″. A long-tailed, robin-sized bird, occurring widely in forest
undergrowth, woodland edges, and brushy areas. Usually feeds on
ground in underbrush, foraging noisily in leaf litter by kicking
backward with both feet. Makes short, low flights from cover
to cover.
Adult males (1) have black hood and upperparts; rich chestnut flanks;
white belly; white corners of tail and white wing patches visible in
flight; eyes red. Adult females (2) similar but with dark brown hood
and upperparts; less white in tail. Florida birds have white eyes.
Western birds ("Spotted" Towhee) similar to eastern but have two
white wing bars, white spots on shoulders. Immatures all resemble
adults by first winter.
Song a distinctive *drink your teeeeeeee*; call note *che-wink*.

Similar Species
American Robin larger, with gray back, slender yellow bill; breast
and flanks rufous.

Range
Breeds from s. British Columbia east to s. Maine and south to
California, n. Oklahoma, e. Louisiana, and Florida. Winters from
s. British Columbia and s. Massachusetts.

400

Dark-eyed Junco
Junco hyemalis

Sparrow-sized
Residential Areas; Forest
habitats; Brushy Open Country

Field Marks
6¼". A widely distributed and common sparrow with several forms in North America. All easily identified by white outer tail feathers in flight, pink bill, white belly, dark eyes. Most widespread is eastern "Slate-colored" Junco. Nests in boreal and mountain forests and edges; in migration and winter, flocks in a variety of habitats, including open woods, brushy areas, edges, and along roadsides. Feeds on ground, primarily on insects and seeds.
Adult male "Slate-colored" (1, 3) has gray hood and back; gray tail with white outer tail feathers conspicuous in flight; belly white; bill pink; eyes dark. Adult females (2) similar but with brownish hood, back, and tail. Juveniles resemble females but streaked above and below, with darker bill; similar to adults by first fall. Males of western form ("Oregon" Junco), rare winter visitors in East, have black head, brown back.
Song a simple, musical trill, usually on one pitch; call notes include a loud *smack*.

Range
Breeds from Alaska east to Newfoundland and south in mountains to California and n. Georgia. Winters throughout United States except s. Texas and s. Florida.

Pine Siskin
Carduelis pinus

Sparrow-sized
Boreal Forest; Thickets; Other
forest habitats

Field Marks
5″. A small, streaked finch, widespread in coniferous and mixed coniferous-deciduous woods in summer; in migration and winter also found in fields, thickets, and backyards. Sharp, pointed bill (thinner than bills of most other finches) distinctive at close range. Like many northern finches, an erratic wanderer, especially in winter when may irrupt south of usual range. Gregarious: often with American Goldfinches and Common Redpolls in winter. Visits feeders. Forages on ground and in trees, eating mostly seeds.
Adults (1, 2) brown above with dark streaks; wings dark with yellow band across base of primaries; tail dark with yellow base (yellow in wings and tail often visible only in flight); underparts dull white with dark streaks. Yellow-washed juveniles resemble adult by first fall. Distinctive, buzzy, up-slurred *sswreeee* notes often interspersed with lively, jumbled song; also sweet, up-slurred *swee* notes.

Similar Species
Immature Red Crossbill larger; mandibles crossed; no yellow in plumage. Juvenile Common Redpoll lacks yellow in wings and tail. See female Purple Finch, House Finch.

Range
Breeds from s. Alaska to Newfoundland, in mountains to California and Texas; in East, to Great Lakes and n. New England. Winters from s. Alaska across s. Canada throughout most of United States.

402

Common Redpoll
Carduelis flammea

Sparrow-sized
Brushy Open Country

Field Marks
5¼″. A small finch of far northern forests and tundra; in migration and
winter also in weedy fields, scrub, and open woods. Often gregarious.
Like many northern finches, irrupts far south of usual range in some
winters, and forms flocks with goldfinches and Pine Siskins. Visits
feeders. Forages in trees and on ground, mostly taking seeds.
Plumage variable. Adult males (2) brown above with dark-streaked
back and rump (sometimes less streaked, quite pale); forehead and
crown patch red; chin black; breast pale to bright pink (sometimes
extending down flanks); belly whitish with flanks and undertail
coverts streaked brown. Winter males less pink. Females (1, 3)
similar to males but pink on breast replaced by gray or pale buff.
Immature similar to female but buffier; lacks red crown patch until
first fall.
Flight call a dry *chit-chit-chit*; also up-slurred, questioning *sueweee*
notes; song mixes these two calls randomly with dry trills.

Similar Species
Hoary Redpoll, *Carduelis hornemanni*, rare winter visitor from
Arctic, very similar but has stubbier bill; usually paler, with pure
white rump.

Range
Breeds from Alaska east across n. Canada. Winters south to central
United States.

403

Purple Finch
Carpodacus purpureus

Sparrow-sized
Forest habitats; Residential Areas
and Parks

Field Marks
6″. A sparrow-sized finch, widespread and common in coniferous and mixed woods, second growth, parks, and suburbs. Often visits feeders in winter. Frequently in small flocks. Eats mostly seeds.
Adult males (1) mostly raspberry-red above, brightest on head and rump; wings and tail brown; face with brown ear patch; raspberry-red below on throat and breast; belly white; flanks inconspicuously streaked with brown; undertail coverts white. Adult females (2), immatures, and first-year males brown above, whitish below with brown streaks; show dark ear patch, light eyebrow and cheek stripe. Adult females have white undertail coverts.
Song a lively, rich warble of slurred notes; calls include a soft *pik*.

Similar Species
Male House Finch red only on breast, eyebrow, and rump; streaked on belly and flanks; female lacks dark ear patch, light eyebrow and cheek stripe. Female Rose-breasted Grosbeak larger, more finely streaked; large, pale bill.

Range
Breeds from British Columbia across Canada, south in the mountains to s. California, to north-central and ne. United States, and in the Appalachians to West Virginia. Winters from s. Canada south to Mexico and the Gulf Coast; absent from Rocky Mountains and Great Basin.

404

House Finch
Carpodacus mexicanus

Sparrow-sized
Residential Areas and Parks;
Urban Areas; Open Country

Field Marks
6″. A sparrow-sized finch, widespread and abundant in a variety of
habitats, especially urban and suburban areas. Expanding in the East,
where introduced. Sociable. Often visits feeders. Diet consists
primarily of seeds.
Adult males (1) mostly brown with red (sometimes orange or yellow)
eyebrow, rump, throat, and breast; paler below, streaked on flanks
and belly. Adult females (2) gray-brown above, whitish below with
dark streaks; face unmarked. Immatures resemble adults by first fall.
Warbled song similar to Purple Finch's but usually not as rich, with
burry notes at end; call a bright *kweat* or *weet*.

Similar Species
Male Purple Finch has more red on upperparts; streaked below only
on flanks; female has dark ear patch, light eyebrow and cheek stripe.
See Pine Siskin.

Range
Resident from s. British Columbia east to w. Nebraska and south to
Mexico and w. Texas. Introduced in New York, has spread to se.
Canada, w. Michigan, and south to central Georgia.

Seaside Sparrow
Ammodramus maritimus

Sparrow-sized
Salt Marshes

Field Marks
6″. A locally common, dark, chunky, large-billed sparrow of Atlantic and Gulf coast salt marshes. Forages along muddy-bottomed tidal creeks lined with rank cordgrass; flies only when pressed. Often lives in loose colonies in association with Sharp-tailed Sparrows.
Adults (1) olive-gray above with short yellow stripe in front of the eye; underparts whitish with blurry grayish stripes; throat white; have white streak along jaw. Juveniles in late summer are browner, with blackish streaks on back; underparts streaked and washed with buff.
Song a buzzy *tuptup zhe-eeeeeeee*, accent on *zhe*; resembles Red-winged Blackbird's song pattern. Frequently sings at night.

Similar Species
Sharp-tailed Sparrow smaller and browner, with distinctive ochre face pattern and well-defined back stripes.

Range
Breeds along Atlantic and Gulf coasts from Massachusetts to Texas; absent from s. Florida. Winters from Virginia southward.

Sharp-tailed Sparrow
Ammodramus caudacutus

Sparrow-sized
Freshwater Marshes; Salt
Marshes

Field Marks
5¼″. A locally common sparrow of Atlantic Coast salt marshes and interior freshwater marshes; considerable geographic plumage variation. Often travels by running mouselike through marsh grass and along muddy tidal creek edges. Flies weakly over tops of marsh grass; appears stubby-tailed in flight.
Adults (1, 2) dark brown above, more or less striped with white or gray; crown gray or dark brown, nape gray, ear patch gray bordered by ochre-yellow eyebrow and face; underparts whitish, breast buffy and streaked (inland birds very buffy, with few streaks); tail short, feathers pointed. Juveniles very buffy overall; show semblance of adult face pattern.
Song a short, insectlike *te-sheeeeeee*; introductory note often inaudible; occasionally sings at night.

Similar Species
Adult Seaside Sparrow larger, grayer, with yellow in front of eye; juvenile Seaside resembles juvenile Sharp-tailed, but larger-billed, more striped on crown, throat whiter. Le Conte's Sparrow has white median crown stripe and black side stripes.

Range
Breeds locally from s. MacKenzie east to nw. Minnesota and along Atlantic Coast from Nova Scotia to North Carolina. Winters along Gulf and southern Atlantic coasts.

407

Le Conte's Sparrow
Ammodramus leconteii

Sparrow-sized
Open Country and Grasslands;
Freshwater Marshes

Field Marks
5″. A fairly colorful sparrow with a flatheaded appearance; inhabits
grassy areas, bogs, and marshes, uttering its song from exposed
perches. Unlike most sparrows, it often runs, rather than flies, when
alarmed.
Adults (1) brownish above, with a warm buff wash on face and breast;
ear patch gray; crown brown, with bold white line running from
forehead to nape. Juveniles similar to adults but paler overall; at close
range, note rust on wings.
Song a thin, insectlike hiss or buzz in two syllables: *tzzt-zzt.*

Similar Species
Sharp-tailed Sparrow has solid crown. Baird's Sparrow has streaked
breast, without buff wash. Henslow's Sparrow has olive-colored face.
Adult Grasshopper Sparrow has no streaking on buff underparts;
juvenile lacks gray ear patch.

Range
Breeds from central Canada south to n. Montana and n. Michigan.
Winters from e. Oklahoma east to South Carolina and south to Gulf
Coast and central Florida.

Grasshopper Sparrow
Ammodramus savannarum

Sparrow-sized
Open Country and Grasslands

Field Marks
5″. A small, chunky, flatheaded sparrow, widespread but often
spottily distributed. Unless singing, usually well concealed in grassy
habitat. Favors prairies, savannas, and old fields, usually with patches
of bare ground between grass clumps.
Adult (1, 2) dark brown above with buff or whitish streaks and
edgings, and variable amounts of rust; crown dark with pale buff or
whitish central stripe; face and breast pale, plain buff to buff-brown,
without streaks; lores yellowish; belly white; tail dark, fairly short.
Immature resembles adult, with lightly streaked breast in first-
fall individuals.
Song a high-pitched, insectlike buzz, usually preceded by two to three
soft *pip* notes.

Similar Species
Immature resembles Baird's Sparrow but lacks latter's dark double
mustache stripes. Henslow's Sparrow has olive head, rusty wings.

Range
Breeds from s. Alberta east to central New England and south to
California, central Texas, and Virginia. Winters from central
California to North Carolina and south.

Baird's Sparrow
Ammodramus bairdii

Sparrow-sized
Open Country and Grasslands

Field Marks
5½″. A small, inconspicuous sparrow of northern short-grass prairies; winters mostly south of United States in weedy fields and grasslands. Formerly more widespread.

Adults (1) dark brown above with buff or whitish streaks and edgings, variably flecked with rust; central crown stripe buff with dark borders; nape buff with dark streaks; face buff, marked by two dark mustache streaks, and dark ear spots that may join upper mustache streak to form dark face triangle; underparts white, usually with buff wash on breast and necklace of dark streaks. Immature duller with less obvious crown stripe.

Song a soft, high-pitched, sweet trill preceded by two to four soft *pip* notes, reminiscent of Grasshopper Sparrow's song, but not buzzy.

Similar Species
Savannah Sparrow not as chunky; has smaller, browner (not buffy) head and smaller bill; often has breast spot. Immature Grasshopper Sparrow lacks double mustache streaks. Henslow's Sparrow has olive head, rusty wings.

Range
Breeds from se. Alberta east to sw. Manitoba and south to South Dakota. Winters from se. Arizona to central Texas, south to Mexico.

Savannah Sparrow
Passerculus sandwichensis

Sparrow-sized
Open Country and Grasslands;
Coastal Dunes

Field Marks
5½". A small sparrow; one of the most widespread North American birds. Tolerates a wide range of climatic conditions and habitats, from tundra to fields, dunes, and marshes.

Variable. Typical adults (2) sandy brown above with dusky streaks; pale central crown stripe bordered by dark stripes; dark double mustache streaks separated by whitish area (upper streak often joins dark ear spot and eyeline to form dark face triangle); eyebrow pale, usually yellowish from bill to eye; white below with dusky streaks on breast and sides; some birds have dark central breast spot. Pale "Ipswich" Sparrow (1) winters on Atlantic Coast; less distinctly marked, especially on back, which may lack streaks; usually retain face pattern. Darkest birds more densely streaked.

Song a thin, insectlike *chip-chip-chipa-bzzzz-bzzzz* (up to eight chips, accelerating; second buzzy trill lower in pitch). Thin *seep* call.

Similar Species
Lincoln's Sparrow has buff mustache and breast. Song Sparrow usually darker, has longer tail; lacks yellow eyebrow. See Baird's and Vesper sparrows.

Range
Breeds across Canada, south to California, central Arizona, Missouri, and New Jersey. Winters from sw. British Columbia to central Arizona across s. United States and along southern Atlantic Coast.

411

Rufous-crowned Sparrow
Aimophila ruficeps

Sparrow-sized
Brushy Open Country; Rocky
Slopes

Field Marks
6″. A long-tailed sparrow with distinctive combination of rusty crown and single black mustache streak. Favors brushy open country and rocky slopes, usually staying concealed near ground.
Adults (1) have rusty crown; gray back with rusty streaks; gray face (palest on eyebrow and above mustache) with rusty eyeline and black mustache streak; wings and long tail dark; gray below. Immature duller than adult, with fine streaks on crown and breast; mustache often obscure; some with indistinct buff wing bars.
Song reminiscent of House Wren's, but not as rich, with notes less slurred.

Similar Species
Other similar sparrows either have double mustache streak or lack mustache streak.

Range
Resident from central California, se. Colorado, Oklahoma, and w. Arkansas southward.

Cassin's Sparrow
Aimophila cassinii

Sparrow-sized
Texas: Grasslands

Field Marks
6". A plain sparrow of dry, open grasslands. Breeding males are
conspicuous when singing in flight; otherwise a rather secretive
species. Feeds on ground, primarily on caterpillars, beetles,
and seeds.
Adults (1) gray-brown above, finely streaked; light eye-ring; long,
rounded tail dark, narrowly tipped with white on outer feathers;
underparts pale gray, usually with a few dark streaks on flanks.
Immatures similar to adults but streaked below.
Song a sweet, musical trill: two short notes (not always heard)
followed by a long, slurred trill, then two short notes dropping
in pitch.

Similar Species
Botteri's Sparrow, *Aimophila botterii*, occurs in tall, dense grasslands
in southern Texas; browner, less streaked above, with longer, more
slender bill; different song: an accelerating series of *chip* notes.

Range
Breeds from s. Arizona east to sw. Kansas and south to s. New
Mexico and s. Texas. Winters from extreme s. New Mexico and
s. Texas into Mexico.

Henslow's Sparrow
Ammodramus henslowii

Sparrow-sized
Brushy Open Country

Field Marks

5″. A flatheaded sparrow of skulking habits; lives in grassy meadows and weedy areas, often with some shrubby vegetation. Prefers to run through cover rather than to fly from intruders. Perches atop weeds and grass stalks to sing.

Adults (2, 3) have large bill, dull olive-colored head with dark brown stripe on crown, dark spot behind ear; upperparts rust or brownish, with whitish streaks, some heavy dark brown spots on shoulders and back; folded wing shows rusty; breast and sides dull buff, with dark brown spots and streaks. Juveniles (1) similar, with few streaks on underparts.

Song a sneezelike *tsee-lik*.

Similar Species

See juvenile Le Conte's Sparrow. Baird's Sparrow has small patch of dull orange on crown and nape; longer tail; looks smaller-billed.

Range

Breeds from s. Minnesota east to central New York and south to e. Kansas and n. North Carolina. Winters from e. Texas to South Carolina and n. Florida.

Vesper Sparrow
Pooecetes gramineus

Sparrow-sized
Open Country and Grasslands;
Coastal Dunes

Field Marks
6¼″. A widespread and fairly common sparrow of grasslands and open country. Male sings from elevated perch; otherwise usually seen on ground. White outer tail feathers conspicuous when flushed.
Adults (1–3) pale grayish-brown above with dark streaks and white eye-ring; indistinct chestnut patch on wings; dark tail with white outer tail feathers conspicuous in flight; underparts whitish with dark streaks; no central breast spot. Immature similar to adult.
Song suggestive of the Song Sparrow's, but throatier; two slurred introductory notes followed by series of short, melodious trills.

Similar Species
Pipits have thinner bills; walk instead of hop. Immature Lark Sparrow has dull but distinct face pattern. Chestnut-collared and McCown's longspurs show more white in tail. Smith's Longspur buff below. Lapland Longspur more rusty on wing. Female Lark Bunting has buff or whitish wing patches.

Range
Breeds from central British Columbia east to Nova Scotia and south to central California, central Texas, and w. North Carolina. Winters from central California to coastal New Jersey, south to Mexico, the Gulf Coast, and central Florida.

Lincoln's Sparrow
Melospiza lincolnii

Sparrow-sized
Thickets; Brushy Open Country;
Northern Wooded Swamps

Field Marks
5¾". A widespread but usually secretive sparrow. Occurs in summer
in northern and mountain bogs and wet meadows; in migration and
winter, primarily in thickets, scrub, weedy fields, and forest edges.
Resembles the closely related Song Sparrow but more slender; tail
shorter; face grayer; breast buff with much finer streaks. Often skulks
in underbrush.
Adults (1, 2) primarily brownish above with streaked back; throat
and belly white; breast buff with fine dark streaks (usually without
central breast spot); eyebrow and sides of neck gray; cheek brownish
with buff mustache; narrow eye-ring white. Immature similar to adult
but with duller streaking; may show spotted throat.
Song, suggestive of House Wren's, a trill that starts low, rises, then
drops.

Similar Species
Song Sparrow plumper, longer-tailed, with less gray on face; whitish
below; breast has much finer streaks and central breast spot.
Immature Swamp Sparrow lacks spotted throat.

Range
Breeds from nw. Alaska east to Newfoundland, and south in the
mountains to s. California and n. New Mexico; in the East, to Great
Lakes region and n. New England. Winters from Arizona east to
w. Mississippi, south to Mexico and Gulf Coast; rarely in s. Florida.

416

Song Sparrow
Melospiza melodia

Sparrow-sized
Thickets; Brushy Open Country;
Residential Areas and Parks

Field Marks
6¼". One of most familiar and widespread North American songbirds.
Highly variable, primarily in West. Habitats vary from weedy fields
and brushy areas to open woods, backyards and gardens, desert
scrub, and salt marshes. Has distinctive habit of pumping long,
rounded tail in flight.
Adults (1, 2) brown above with dark streaks on back and gray streak
on crown; tail dark, with rusty tinge; face with broad gray eyebrow
and dark mustache; throat white; underparts whitish with brown
streaks and dark central breast spot. Juvenile buffier below, more
finely streaked, often without central breast spot; resembles adult by
first winter.
Song variable; usually three to four clear introductory notes followed
by a musical trill, dropping in pitch. Call note a loud *chimp.*

Similar Species
Lincoln's Sparrow more gray on face; buffier and more finely streaked
below; no central breast spot. Swamp Sparrow with less streaking
below; more gray on face; more rust-brown above, especially on
wings. See Savannah and Fox sparrows.

Range
Breeds from s. Alaska east to w. Newfoundland and south to New
Mexico, ne. Kansas, n. Arkansas, and n. Georgia. Winters from
s. Alaska through most of U.S. and n. Mexico.

Clay-colored Sparrow
Spizella pallida

Sparrow-sized
Brushy Open Country

Field Marks
5½". A small, slim sparrow, locally common in brushy open country and shrubby areas. Forages on ground or low shrubs, feeding heavily on seeds.
Breeding adults (1, 2) primarily buff-brown above, marked with some rust tones and dark streaks; wings with two whitish or buff wing bars; rump unstreaked buff-brown; collar gray; crown dark-streaked with buff central stripe; eyebrow buff-gray; cheek patch buff-brown outlined with dark; throat white with faint dark mustache; below whitish with buff sides; bill pale with dark tip. Winter adults similar but rustier above, more buff below. Immatures resemble winter adults but crown stripe indistinct. Briefly held juvenal plumage like immature's but streaked on breast and sides.
Song a series of (usually two to four) insectlike *bzzzz* notes.

Similar Species
Chipping Sparrow has gray rump in all but juvenal plumage; brown cheek only bordered with dark above.

Range
Breeds from ne. British Columbia east to Ontario and south to sw. Colorado and central Michigan. Winters from s. Texas to Mexico.

418

Chipping Sparrow
Spizella passerina

Sparrow-sized
Forest habitats; Brushy Open
Country; Residential Areas

Field Marks
5½″. A small, slim sparrow, widespread in woodlands, forest edges, and backyards; in migration and winter also occurs in brushy areas and along roadsides. Quite tame; often nests close to houses. Breeding adults (1, 2) have distinctive rust-colored cap on head, white eyebrow, black line through eye, dark bill; back reddish-brown with dark streaks; two white wing bars; cheeks, collar, rump, and underparts plain gray. Winter adults have duller, streaked crown; face and eyebrow brownish with dark eyeline; lighter bill. First-winter birds like winter adults, but streaked crown and wing bars are buff; grayish-buff below. Juvenal plumage (sometimes held into fall) streaked dark above, including rump, and on grayish-buff underparts. Young birds have pinkish bill.
Song a long, monotone trill of *chip* notes.

Similar Species
Winter adult and immature Clay-colored Sparrows have brown rump and dark-bordered cheek patch. Juvenile Clay-colored has stronger face pattern; more buff-colored below.

Range
Breeds across Canada from Yukon east to Newfoundland and south to California and Gulf Coast. Winters from California, Tennessee, and Maryland southward; rarely from Oregon, Great Lakes region, and New York.

American Tree Sparrow
Spizella arborea

Sparrow-sized
Open Country

Field Marks
6¼". A far northern nester; widely distributed in weedy fields and brushy areas across United States in winter. Easily identified: combination of central breast spot and two white wing bars a reliable field mark among North American sparrows. Usually feeds on ground, almost exclusively on seeds.

Adults (1, 2) have gray head with distinctive bright rust-colored cap and eyeline; back rust-brown, streaked; two white wing bars; bill two-toned (dark above, light below); underparts pale gray with dark central breast spot; rusty patches on sides of breast. Juvenile has streaked breast and crown; resembles adult by first fall.

Song a syncopated series of clear, whistled notes, often doubled or tripled; highest-pitched at beginning; flock call a distinctive *teedle eet*.

Similar Species
Chipping and Field sparrows easily told by lack of central breast spot in all plumages.

Range
Breeds from n. Alaska east to n. Labrador. Winters from s. Canada to n. California, central Texas, and South Carolina.

Field Sparrow
Spizella pusilla

Sparrow-sized
Brushy Open Country

Field Marks
5¾". A locally common, slim, long-tailed sparrow of overgrown pastures, old fields, fencerows, and open deciduous forest edges and clearings; decreasing in some areas due to habitat change. Sings persistently late into summer. Small flocks often occur in tangles and weed fields in winter.

Adults (1, 2) have rusty crown, gray face, white eye-ring, two white wing bars, and brown back, streaked with dark; bill bright pink; underparts grayish white, with buff wash across breast and flanks; legs pink. Juveniles in late summer duller, with light streaks on breast and sides.

Song a plaintive, clear, whistled *teeu teeu teeu tee tee tee wee wee wee wee wee*; slow at start, accelerating to trill at end. Call note a thin *seeu* and sharp *chip*.

Similar Species
Chipping Sparrow has dark bill, black eyeline, white eyestripe, and gray underparts. American Tree Sparrow has dark upper mandible, yellow lower mandible, and dark breast spot.

Range
Breeds from e. Montana east to s. Maine and south to w. Colorado and Georgia; rarely to n. Florida. Winters from s. Nebraska to e. Massachusetts south to Texas and central Florida.

421

Bachman's Sparrow
Aimophila aestivalis

Sparrow-sized
Southeastern Pine Forest;
Thickets; Brushy Open Country

Field Marks
6″. A rather nondescript sparrow of southern pine forests, thickets, and brushy areas in open country. Forages on the ground for beetles, crickets, and other insects; perches atop low shrubs or weed stalks to sing. When flushed, will usually run for cover.
Adults (1) reddish-brown above, with red crown, buff eyebrow, and reddish line through eye; underparts clear, unmarked buff, palest on belly; tail long, somewhat rounded. Juveniles similar with some streaks on upper breast and sides.
Song a loud, sweet, whistled note, followed by a slower trill on a different pitch; repeated with variations.

Similar Species
Swamp Sparrow has gray breast and face; immature lacks streaking on breast and sides. Grasshopper Sparrow has dark crown with buff or yellow stripes; lacks reddish on upperparts; immature has dark crown, paler underparts. Botteri's Sparrow nearly identical, but ranges do not overlap.

Range
Breeds from ne. Texas, s. Ohio, and Maryland south to Gulf Coast and central Florida. Winters in southern parts of breeding range.

Lark Sparrow
Chondestes grammacus

Sparrow-sized
Open Country and Grasslands;
Residential Areas and Parks

Field Marks
6½". A large, distinctive sparrow of open lands, cultivated areas, and forest edges. Adult's chestnut-and-white-striped head and dark tail with white corners unmistakable. Gregarious; often in large flocks in winter. Primarily a ground-dweller and seed-eater.

Adults (1) have chestnut-and-white-striped head; back and wings brownish; dark, rounded tail with white corners conspicuous in flight; underparts white, with dark breast spot. Immature has streaking on crown and breast; lacks breast spot; duller face pattern; tail like adult's.

Song a lively jumble of trills interspersed with whistled and burry notes.

Similar Species
Vesper Sparrow also shows white on outer tail, but lacks distinctive face pattern; more streaked above and below.

Range
Breeds from British Columbia east to w. Ohio and south to Mexico, Texas, and n. Alabama. Winters from central California south to central Texas and Mexico. In fall, occurs in small numbers along Atlantic Coast.

Eurasian Tree Sparrow
Passer montanus

Sparrow-sized
Missouri and Illinois: Open
Country

Field Marks
6″. An introduced, energetic sparrow with a brown head; found only near St. Louis, Missouri, where it inhabits open country, farmland, and residential areas. Forms large flocks of as many as 100 birds in winter.
Adults (1) have warm brown head, white cheeks, black throat and cheek patch; upperparts brown, underparts pale brown to buff. Immatures similar, but black on throat and cheek replaced by gray; some darker brown mottling on head and upperparts.
Song a repetitive chirp; also chatters.

Similar Species
Male House Sparrow in breeding plumage has gray crown; cheek lacks black patch over ears; markings less neat, especially black throat.

Range
Established around St. Louis, Missouri; has spread into w. Illinois.

House Sparrow
Passer domesticus

Sparrow-sized
Residential Areas and Parks;
Urban Areas; Open Country and
Grasslands

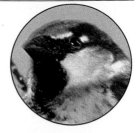

Field Marks

6¼″. The familiar "sparrow," abundant in cities, towns, and on farms across the continent. Introduced from Old World. Has shorter legs, thicker bill than native sparrows. Noisy, highly sociable, feisty. Often nests on buildings.

Breeding males (1) brown above with streaked, chestnut nape; crown gray; cheek whitish; bill black; wing bar white; grayish below with prominent black bib. Females (3) brownish, paler below than above; back streaked; eyebrow pale. Winter males (2) have black only on chin; bill yellowish. Immatures similar to females. City birds usually dirtier and duller than rural birds.

No true song; chirps and cheeps repeatedly.

Similar Species

Habitat, voice, and behavior distinguish this species from native sparrows. Male Eurasian Tree Sparrow has much neater markings on plumage.

Range

Introduced and established from central Canada south throughout the United States.

Black-throated Sparrow
Amphispiza bilineata

Sparrow-sized
Southern Texas: Brushy Open
Country

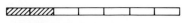

Field Marks
5½″. A fairly common bird of desert scrub and brushy open country; tolerant of extremely hot and dry conditions. Striking black-and-white face and black throat of adult unmistakable. Usually seen on ground, where feeds on insects and seeds.

Adults (1–3) primarily a plain brownish-gray above, with blackish cap and blackish tail with white outer tail feathers; broad white eyebrow and whisker stripe contrast boldly with blackish sides of face and lores; throat black; rest of underparts whitish or pale grayish. Juveniles browner, streaked above and below; lack black throat but show dull adult face pattern; resemble adults by first winter.

Song composed of two short, rising notes followed by a rapid trill.

Range
Breeds from ne. California east to sw. Wyoming and se. Colorado, south to Mexico. Winters from s. Arizona into Mexico.

Harris' Sparrow
Zonotrichia querula

Sparrow-sized
Thickets; Brushy Open Country;
Boreal Forest

Field Marks
7½". A large, pink-billed, distinctively patterned sparrow that winters chiefly in brushy areas of the Great Plains, although wanderers turn up throughout our range. Forages on the ground, scratching for insects and seeds.

Adults (3) have pink bill, black mask on face and throat, gray cheeks; underparts white with black spots and streaks; back striped with dark brown. Immatures (1, 2) similar, but have buff head; sometimes show traces of adult's black mask.

Song a series of two or three high notes; series repeats on same or different pitch.

Similar Species
Winter-plumage Lapland Longspur similar to immature, but with black mark on buff cheek, more reddish-brown in wings, and white outer tail feathers; tail also shorter; habitat different.

Range
Breeds in north-central Canada. Winters from s. Minnesota south along eastern Great Plains to south-central Texas; occasionally in interior West and Southwest, rarely along Pacific and Atlantic coasts.

427

Fox Sparrow
Passerella iliaca

Sparrow-sized
Thickets; Brushy Open Country;
Other forest habitats

Field Marks
7″. One of the most variable North American songbirds, and one of
our largest sparrows. Typically found in dense cover in thickets,
forest edges, and second growth.
Adults (1) rust-brown above, brightest on rump and tail, with gray
eyebrow, collar, and back streaks; whitish below with small rusty
"arrowheads" forming heavy streaks and breast spot. Immatures
resemble adults.
Song reminiscent of Song Sparrow's, but richer and less buzzy;
usually begins with sweet warbled notes, ends with a trill or burry
phrase.

Similar Species
Large size, rust-brown coloring distinctive. Song Sparrows show
prominent dark mustache.

Range
Breeds from n. Alaska east to Newfoundland and south to s.
California, Nevada, and Colorado. Winters from s. British Columbia
south to s. California, and from s. Kansas east to New Jersey and
south to Gulf Coast.

Swamp Sparrow
Melospiza georgiana

Sparrow-sized
Freshwater Marshes

Field Marks
5¾". A fairly common, dark, shy sparrow of freshwater marshes,
brushy wet meadows, bogs, and swamps; weed fields and open brook
edges in winter. Forages low amid dense wetland vegetation; may be
induced into view by sharply kissing back of hand.
Breeding adults (1) dark brown above, streaked with black; crown
reddish and wings rusty; throat and belly white; breast and sides
gray, with fine dusky streaks; flanks brownish. Winter adults (2) and
immatures buffier, with dark-streaked crown, pale median stripe.
Song a loud, metallic, even-pitched trill: *weet-weet-weet-weet-weet*.
Call note a sharp *chip*.

Similar Species
Chipping Sparrow found in different habitat; has black eyeline, whiter
eyebrow. American Tree and Field sparrows have conspicuous wing
bars. In northern bogs in summer, juvenile Swamp Sparrows
resemble Lincoln's Sparrows, but are darker, with blacker crown,
redder wings, and usually unspotted throat.

Range
Breeds from ne. British Columbia east to Newfoundland and south
to n. Missouri and Maryland. Winters from e. Nebraska and
Massachusetts south to Texas and Florida.

White-crowned Sparrow
Zonotrichia leucophrys

Sparrow-sized
Thickets; Brushy Open Country;
Other forest habitats

Field Marks
7″. A large and distinctive sparrow, widespread in woodland edges, thickets, brushy open country, and roadsides; often around houses. Usually feeds on ground, primarily on seeds. Frequently seen in flocks in winter.
Adults (1, 2) have striking black-and-white-striped crown; brownish, streaked back and wings, latter with two white wing bars; face, nape, and breast gray; whitish chin and belly; pale bill (varies from yellow to pink). Immatures (3) similar but with brown and buff head stripes. Song, suggestive of White-throated Sparrow's, varies geographically; usually a few introductory notes followed by a simple trill.

Similar Species
White-throated Sparrow stockier, less erect, with sharply defined white throat; lores usually yellow; bill dark; posture less erect.

Range
Breeds from n. Alaska east to Labrador and south to south-central California and n. New Mexico. Winters along Pacific Coast from s. Alaska to Mexico and across s. United States to Maryland.

White-throated Sparrow
Zonotrichia albicollis

Sparrow-sized
Thickets; Brushy Open Country;
Boreal Forest; Other forest
habitats

Field Marks
6¾". A large sparrow with erect posture, common in boreal forests
and edges, thickets, and bogs; in migration and winter also occurs in
deciduous woods, brushy areas, and gardens. Usually feeds on
ground, primarily on seeds. Often in flocks in winter.
Adults (1, 3) primarily brown above, with streaked back and two
white wing bars; crown striped with black and white or brown and
tan; lores usually yellow; throat distinctly white; cheeks and breast
gray; belly whitish; bill dark. Immatures (2) usually have crown
streaked with brown and tan (sometimes white); duller throat; may
lack yellow lores; breast may be finely streaked.
Song usually of two clear whistles followed by three triplet whistles;
often rendered as *Old Sam Peabody Peabody Peabody* or *Oh Sweet
Canada Canada Canada*; easily imitated.

Similar Species
White-crowned Sparrow lacks distinct white throat and yellow lores;
has pale bill and still more erect posture.

Range
Breeds from s. Yukon east to Newfoundland and south to central
Michigan, n. Pennsylvania, and Massachusetts. Winters from n.
California to s. New Mexico; and e. Kansas to Massachusetts, south
to central Texas and central Florida.

Golden-crowned Kinglet
Regulus satrapa

Very Small
Boreal Forest; Eastern Deciduous
Forest; Residential Areas and
Parks

Field Marks
4″. A tiny, wing-flicking species of northern coniferous forests, eastern deciduous forests, and parks and suburban areas. Forages actively for insects and larvae among tree branches; sometimes hovers like a hummingbird before twigs.
Adults (1) are grayish-olive above, paler gray on underparts; crown patch bright orange-yellow in males (1, left), paler yellow in females (1, right); bordered with black in both sexes; eyebrow white; wing darker than back, with conspicuous white wing bars; tail short; bill small and stubby.
Call a high, thin lisping *tseee, tseee*, rising in pitch. Song a high-pitched pattern of notes.

Similar Species
Ruby-crowned Kinglet has darker underparts and lacks black on crown and white eyebrow; has different call; red crown patch not always visible in the field. Warblers larger, with larger bills.

Range
Breeds from s. Alaska east to Newfoundland and south in mountains to s. California, to central Minnesota and New England, and in the Appalachians to w. North Carolina. Northern birds withdraw in winter, reaching south to the Gulf Coast and Florida.

Worm-eating Warbler
Helmitheros vermivorus

Sparrow-sized
Eastern Deciduous Forest

Field Marks
5¼". A somewhat inconspicuous inhabitant of dry, wooded hillsides
and forested areas. Spends much time on the ground, foraging slowly
for small moth larvae and other insect prey, but sings from perches
high in trees.
Adults (1) olive-green above, buff below, with four distinct blackish
stripes on buff crown; bill slender and pointed. Immatures resemble
adults.
Song a dry, insectlike, buzzy trill, somewhat like Chipping Sparrow's
song.

Similar Species
Swainson's Warbler similarly colored above, but head and underparts
paler yellowish white, not warm buff; has reddish-brown cap on
crown.

Range
Breeds from e. Iowa to s. New England and south to e. Texas and n.
Georgia. Winters in Mexico and Central America; rarely in s. Florida.

433

Wilson's Warbler
Wilsonia pusilla

Sparrow-sized
Boreal Forest; Thickets; Other
forest habitats

Field Marks
4¾". A small, bright yellow warbler that flits actively about near the ground. Usually encountered in thickets near water in boreal forests and other wooded habitats.

Adults basically yellow above and below; upperparts somewhat greener, underparts somewhat yellower, with no markings in wings or tail. Males (1) have glossy black circular spot on crown. Females may have dark cap, but most are plain olive-green on crown. Dark eyes of both sexes stand out against yellow cheeks and lores, creating "beady-eyed" look.

Song a short chattering trill, dropping in pitch.

Similar Species
Female Hooded Warbler larger, with bold white spots in tail and dark-tinged lores (does not look beady-eyed); some individuals may have black crown, but with different shape, usually extending behind ears. Female Yellow Warbler has yellow wing bars and tail spots, paler crown.

Range
Breeds from n. Alaska east to Newfoundland and south to s. California, n. New Mexico, and n. New England. Winters from s. California and s. Texas to Central America.

Hooded Warbler
Wilsonia citrina

Sparrow-sized
Eastern Deciduous Forest;
Northern Wooded Swamps;
Southern Wooded Swamps;
Thickets

Field Marks
5¼". A fairly common but elusive warbler found in the understory of low, mature deciduous woodlands, wooded swamps, and along watercourses. Forages by actively flycatching in undergrowth; often flicks tail, showing white outer tail feathers.
Adult males (1, right) have black hood surrounding yellow face; underparts yellow; upperparts plain olive-green, with extensive white outer tail feathers. Females (1, left) and immatures lack hood, although some black often present about the head; outer tail feathers white.
Song a loud, rapid *weeta weeta wee-tee-o*, first rising and then falling at end; note a loud *chink*.

Similar Species
Female Wilson's Warbler smaller, with different head pattern and no white in tail. Bachman's Warbler, *Vermivora bachmanii*, very rare summer visitor to southeastern swamps, has black cap and breast patch, but yellow of face not completely surrounded by black; female similar to female Hooded, but has gray on crown, nape, and breast.

Range
Breeds from se. Iowa east to s. New England and south to e. Texas and n. Florida. Winters in Mexico and Central America.

Black-throated Green Warbler
Dendroica virens

Sparrow-sized
Boreal Forest; Eastern Deciduous
Forest

Field Marks
5″. A locally common warbler of mixed and coniferous northern forests; forages high on the outer branches of trees and often sings persistently in the process.
Adult males (2) yellowish-green above with a yellow face, dark eyeline, and two white wing bars; throat black; white underparts striped with black on sides; lower belly has faint yellow wash.
Females (1) and immatures (3) similar, but with less black on throat and sides.
Song a high-pitched, rambling *zee-zee-zee zer zee* or *zer zee zer zer zee*. Note an emphatic *tsep*.

Similar Species
Golden-cheeked Warbler, *Dendroica chrysoparia*, found in juniper woodlands in Texas, where Black-throated Green is rare; has black back, black line through eye, and pure white lower belly; males have black crown.

Range
Breeds from n. Alberta east to Newfoundland, south to central Minnesota, n. New Jersey, and in the mountains to n. Alabama and n. Georgia. Winters in s. Texas and s. Florida; also in Central and South America.

Prairie Warbler
Dendroica discolor

Sparrow-sized
Brushy Open Country; Pine
Barrens; Eastern Deciduous
Forest

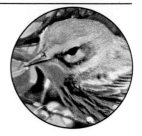

Field Marks
4¾". A small warbler of brushy open country, pine barrens, and mangrove swamps. Regularly pumps tail while foraging in low, brushy locations; often sings through heat of the day.
Adult males (1, 2) olive-green above with two yellowish wing bars and black marks through eye and along jaw; underparts bright yellow, with black streaks on sides of breast and flanks; white spots in tail. Females and immatures (3) duller, especially facial markings and side streaks.
Song a distinctive series of *zee* notes, ascending the scale.

Similar Species
Palm Warbler also pumps tail, but is less yellow below, has more extensive breast streaking, brownish or chestnut cap. Immature Magnolia Warbler has yellow rump and more conspicuous white tail patches.

Range
Breeds from central Missouri and Michigan east to s. Maine and south to e. Texas, n. Louisiana, and Florida. Winters from n. Florida south; rarely along Gulf Coast.

Blue-winged Warbler
Vermivora pinus

Sparrow-sized
Brushy Open Country; Eastern
Deciduous Forest

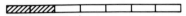

1

Field Marks
4¾". A locally common warbler of brushy open country, woodland edges, and second-growth deciduous woods. Forages in lower tree branches and undergrowth; males often sing from treetops. Range expanding toward northeast and west; displacing Golden-winged Warbler in some areas.

Adults (1) have black eyeline, yellow crown, and yellow underparts; greenish-yellow back contrasts with bluish-gray wings and tail; two white wing bars and white spots in tail. Immatures similar, but with less contrast between crown and back.

Song typically a buzzy insectlike *beee-bzzzz*, second note lower and drawn out; occasionally sings song of Golden-winged Warbler.

Similar Species
"Brewster's" Warbler, hybrid between Blue-winged and Golden-winged warblers, has underparts white or partly white, often has yellow wing bars. Pine Warbler more olive, lacks black eyeline, has streaked flanks.

Range
Breeds from e. Nebraska east to s. Maine, south on coastal plain to ne. Oklahoma, n. Georgia, and n. Virginia. Winters in Central America.

Canada Warbler
Wilsonia canadensis

Sparrow-sized
Eastern Deciduous Forest; Other
forest habitats

Field Marks
5¼". A common, active, flycatching warbler of luxuriant undergrowth
in deciduous and coniferous forest thickets; also in cool maple and
cedar swamps. Usually forages within a few feet of the ground; can be
difficult to observe.

Adult males (1, 2) plain blue-gray above and bright yellow below, with
yellow spectacles, black sideburns, and a necklace of black streaks
across the breast. Females (3) and immatures similar, but with eye-
ring and necklace less distinct.

Song an irregular burst of sharply punctuated notes, none on the same
pitch; often introduced or terminated by a sharp *chip*. Call note a loud
chick.

Similar Species
Kentucky Warbler has similar face pattern but is olive-green above
and lacks black necklace.

Range
Breeds from central Alberta east to Nova Scotia, south to central
Minnesota, n. New Jersey, and in mountains to n. Georgia. Winters in
South America.

Yellow Warbler
Dendroica petechia

Sparrow-sized
Brushy Freshwater Marshes;
Thickets; Residential Areas and
Parks

Field Marks

5″. The yellowest of warblers and a familiar inhabitant of parks,
residential areas, and thickets, usually near water. Especially fond of
streamside willows. Found almost throughout North America, with
several different geographical races varying from deep gold (in the
Florida Keys) to a very pale form of the Southwest.
Adults of both sexes predominantly yellow throughout; olive-yellow
wings and tail, marked by yellow wing bars and tail spots. Males (1)
brighter than females (2), with rust-colored streaks on breast. Streaks
faint or absent in females and immature males. Immature females
much duller than adult, in some areas nearly pale gray.
High-pitched, patterned song begins with three similar notes followed
by a more varied phrase: *sweet-sweet-sweet-sue-so-sweet.*

Similar Species

Female Wilson's Warbler darker on crown, lacks spot in tail. Brighter
variants of Orange-crowned Warbler lack yellow tail spots, show
contrasting pale eyebrow, dark lores, and vague streaks below.

Range

Breeds from Alaska east to Newfoundland and south to s. California,
n. Oklahoma, and n. Georgia; also in Florida Keys. Winters from
Mexico to South America; occasionally in s. Florida, rarely in
s. Arizona.

Prothonotary Warbler
Protonotaria citrea

Sparrow-sized
Southern Wooded Swamps

Field Marks
5½″. A cheerful, bright, orange-yellow songster, and a welcome
harbinger of spring in much of the East. Especially common in
wooded swamps of the Southeast in summer; nests in tree cavities,
unlike most other warblers. Generally seen on or near ground.
Adult males (1, 2) have bright orange-yellow head and breast; back
olive; wings blue-gray; dark tail has large patches of white. Adult
females (3) similar but not as bright on head and breast. Immatures
resemble females. In all plumages, note dark eyes and bill.
Song a sweet, ringing series of *zweet-zweet* notes; male sometimes
gives canarylike flight song. Call note a *tink*.

Similar Species
Blue-winged Warbler has white wing bars and prominent dark gray or
black eyeline; "Brewster's" Warbler (hybrid of Blue-winged and
Golden-winged warblers) has white or yellow wing bars and black
eyeline; in both species, yellow on head and breast paler, with less
orange tone. Yellow Warbler has much more yellow in upperparts; tail
olive-yellow with yellow patches.

Range
Breeds from se. Minnesota east to sw. New York and central
New Jersey, south to e. Texas and central Florida; absent from
Appalachian region. Winters in Central and South America; rarely
in Florida.

441

Pine Warbler
Dendroica pinus

Sparrow-sized
Eastern Deciduous Forest;
Southeastern Pine Forest and
Pine Barrens

Field Marks

5½". A common, large-billed, rather plain warbler of open pine
woodlands and pine barrens. Sluggishly forages by creeping on pine
branches at various levels, often alighting briefly on ground or trunks
of pine trees; occasionally visits suet feeders in winter.

Adults (2, 3) unstreaked olive-green above with two white wing bars
and white tail spots; underparts yellow, with indistinct dark streaks
on breast and sides; undertail coverts white. Immatures (1) brownish
or brownish-olive above, with brownish cheeks; underparts whitish or
ashy, sometimes with yellowish wash; flanks grayish.

Song a musical, ringing trill, slightly variable in pitch; slower than
similar song of Chipping Sparrow.

Similar Species

Chunkier Yellow-throated Vireo has heavier bill, yellow spectacles,
unstreaked underparts, and no tail spots. Immature Bay-breasted and
Blackpoll warblers have streaked backs; immature Cape May Warbler
smaller, with yellowish rump and neck patch, heavy streaks beneath.

Range

Breeds from se. Manitoba east to central Maine and south to central
Texas and Florida. Rare in Mississippi and Ohio river valleys.
Winters from se. Oklahoma to s. New Jersey and south; sparingly
along coast from Massachusetts.

Cape May Warbler
Dendroica tigrina

Sparrow-sized
Boreal Forest

Field Marks
5″. A small, short-tailed, locally common warbler of open, boreal spruce forests; occasionally common as a fall migrant along northern Atlantic Coast and offshore islands. Forages methodically among dense clusters of needles, periodically hovering momentarily at the tips of branches.

Breeding males (3) yellowish-green above with chestnut cheek patch and white wing patch; side of neck and rump yellow; underparts yellow, heavily streaked with black. Breeding females (2) duller, with streaked yellowish-white breast and no chestnut cheek patch. Winter adults (1) and immature males grayer overall; males have less distinct cheek patch. Immature female totally grayish above, heavily streaked below with only a trace of yellowish on breast and rump.

Song a very thin, sibilant *seet seet seet seet*, all on same pitch.

Similar Species
Immature Yellow-rumped Warbler larger and browner, lacks yellow neck patch and has white eyelids. Palm Warbler has dull yellow rump and undertail coverts, pumps tail, and is usually seen near the ground.

Range
Breeds across s. Canada to Nova Scotia, south to North Dakota, Great Lakes region, and n. New England. Winters mainly in the West Indies.

American Redstart
Setophaga ruticilla

Sparrow-sized
Eastern Deciduous Forest;
Thickets; Other forest habitats

Field Marks

5¼". An unmistakable, fluttery warbler. Abundant in deciduous forest understory, woodlots, and thickets. Can be seen on migration well into June, long after other warblers have reached their breeding grounds. Often fans tail, both when perched and in flight, to reveal the colorful pattern.

Adult males (3) glossy black with orange patches in wings, tail, and sides of breast; belly white. Females (1, 2) have same pattern in gray and yellow. First-fall males similar to females, but darker by first spring, with salmon-colored breast patches and some black spots on breast; reach adult plumage by second fall.

Song a series of slow, high-pitched notes of variable pattern, usually ending in a note slurred up or down.

Range

Breeds from se. Alaska east to Labrador and Newfoundland, south to Utah, e. Oklahoma, northern Gulf Coast region, and along Atlantic Coast to Georgia. Winters from s. Texas and s. Florida to South America.

444

Blackburnian Warbler
Dendroica fusca

Sparrow-sized
Forest habitats

Field Marks
5″. A strikingly brilliant warbler, fond of tall hemlocks and dense northern spruce-fir forests. Forages high in the inner canopy of thick evergreens, and thus is difficult to observe. Appears in a variety of habitats during migration, but its high-pitched song often goes undetected.

Adult males (1, 3) unmistakable, with bright orange throat, eyebrow, and crown patch, black back striped with white, and large white wing patch; underparts whitish with black stripes on sides. Females, immatures (2), and males in winter are similar, but orange on throat is replaced by yellow, and two white wing bars are present.

Song a high, thin, wiry series of notes, like *sip sip sip titi tzeeeee*, last note ascending; variable.

Similar Species
Yellow-throated Warbler has unstreaked gray back and white eyebrow stripe.

Range
Breeds from Saskatchewan to Nova Scotia, south to Great Lakes region, s. New England, and in the Appalachians to n. Georgia. Winters in Central and South America.

Palm Warbler
Dendroica palmarum

*Sparrow-sized
Open Country; Thickets; Northern
Wooded Swamps*

Field Marks
5½″. A fairly common warbler of northern bogs in summer and open
woodland edges and fields in winter; one of the earliest and latest
warblers to migrate. Constantly pumps tail while foraging in low
bushes and on the ground.
Breeding adults (1, 2) brownish above with a chestnut crown and
yellowish eyebrow; underparts yellow or grayish-yellow, streaked
with chestnut; undertail coverts yellow. Winter adults (3) and
immatures similar but browner overall, including crown.
Song a buzzy trill; call note a dry *check*.

Similar Species
Prairie Warbler also pumps tail, but lacks chestnut crown, has dusky
jaw stripe, and below has stripes on sides only. Yellow-rumped
Warbler has more extensive bright yellow rump, does not pump tail.

Range
Breeds from s. Mackenzie east to Nova Scotia, south to Great Lakes
region and Maine. Winters from North Carolina south along Atlantic
and Gulf coasts to Texas; occasionally from s. New England.

Yellow-throated Warbler
Dendroica dominica

Sparrow-sized
Eastern Deciduous Forest; Other
forest habitats

Field Marks
5½". An inhabitant of bottomlands, deciduous forests, and wooded
areas with pine, cypress, sycamore, and oak; in South, especially fond
of habitats with plenty of Spanish moss. Tends to stay high in the
trees; sometimes creeps along branches to forage for insects.
Adults (1–3) have contrasting black-and-white head pattern, bright
yellow throat and upper breast; whitish belly and white flanks with
black streaks; upperparts entirely gray except for white wing bars.
Immatures resemble adults. Hybridizes rarely with Northern Parula;
offspring, known as "Sutton's" Warbler, similar to adult Yellow-
throated but with less streaking on flanks and a greenish-yellow patch
in center of breast.
Song a loud, melodic series of clear notes, accelerating as they
descend, ending on one higher note: *teeew-teew-teew-tew-tew-tew-twi*.

Similar Species
Magnolia Warbler has much more extensive yellow on underparts and
more prominent white wing bars.

Range
Breeds from central Missouri south to se. Texas, east to s. New
Jersey and central Florida. Winters from South Carolina along
Atlantic and Gulf coasts to se. Texas; also inland in Florida and in
Central America.

447

Magnolia Warbler
Dendroica magnolia

Sparrow-sized
Boreal Forest; Other forest
habitats

Field Marks
5″. A common warbler found in stands of spruce and fir in the boreal forest; also inhabits mixed coniferous-deciduous forest. Forages actively, high in the center of tree canopy, making it hard to observe. Often fans tail.
Breeding males (2) blackish above with white eyebrow, black mask, white wing patch, yellow rump, and large white tail patches; throat and underparts yellow, breast and sides streaked with black. Females (1) duller, less black below and with less white in wing. Winter adults (3) and immatures grayish-olive above; underparts yellow, with faint side streaks and narrow, pale gray breast band.
Song a short, rising, hurried *wee-o wee-o wee-chy*; call note a flat *tlep*.

Similar Species
Breeding Yellow-rumped Warbler has white underparts. Cape May Warbler has chestnut cheeks, yellow patch on neck. Immature Prairie Warbler lacks white tail band and yellow rump, pumps tail, and has dusky jaw stripe.

Range
Breeds across Canada, south to Great Lakes region, New England, and locally in mountains to Pennsylvania and Virginia. Winters in Central America.

Kirtland's Warbler
Dendroica kirtlandii

Sparrow-sized
Michigan: Pine Barrens

Field Marks
5¾″. A very rare blue-gray warbler; breeds only in small area of pine barrens in Lower Michigan, and seldom sighted in migration. Often wags or bobs tail as it forages for insects along lower branches of scrubby growth.
Breeding males (1) dark blue-gray above, pale yellow below, with some streaking on flanks; face black, with distinct but incomplete white eye-ring. Females (2) similar but paler, with no black on face. Immatures resemble females.
Song a relatively low, loud series of distinct bubbling notes.

Similar Species
Magnolia Warbler has yellow rump and yellow spots in tail. Yellow-throated Warbler has less extensive yellow on underparts.

Range
Breeds locally in Lower Michigan; occasionally strays to Wisconsin and s. Canada. Winters in Bahamas.

Yellow-rumped Warbler
Dendroica coronata

Sparrow-sized
Boreal Forest; Thickets; Other
forest habitats; Coastal Dunes

Field Marks
5½". A thin-billed, sparrow-sized bird. The eastern form, commonly called "Myrtle" Warbler, is the most common and widespread winter warbler in the East; replaced in most of West by "Audubon's" Warbler. Nests in boreal and mixed forests, but in colder months frequents brushy thickets, hedgerows, and coastal dunes. All plumages characterized by a prominent yellow rump patch. Breeding males (3) patterned in gray, black, and white, with yellow on crown and sides of breast as well as rump; white eyebrow and throat set off prominent black cheeks. Spring females duller and browner, but show same basic pattern. Winter birds (1, 2) and immatures duller and browner still, with much less yellow on sides of breast, none in crown, and dark brown cheeks.
Song a variable rambling trill or warble. Call note a loud *chek*.

Similar Species
Spring Magnolia Warbler has a yellow throat, white tail patches, and more prominent white in wing. Sparrows lack thin bill and yellow rump. Palm Warbler forages on the ground and pumps tail.

Range
"Myrtle" breeds from Alaska to Newfoundland, south to Great Lakes and New York. Winters mostly in se. United States; occasionally to Great Lakes. "Audubon's" breeds from British Columbia to Central America.

Chestnut-sided Warbler
Dendroica pensylvanica

Sparrow-sized
Brushy Open Country

Field Marks
5″. A small, locally common warbler, found in brushy second growth and along forest edges and power lines. Forages in lower story of trees, where it is easily observed; often droops wings and cocks tail when perched.

Breeding adults (1, 3) have yellow crown, white underparts with chestnut sides, yellow wing bars, and back streaked with yellow-green and black. Immatures (2) yellow-green above with yellow wing bars, a white eye-ring, and whitish underparts.

Song an emphatic, whistled *please please pleased ta meecha*, rising, then dropping at end; also a second, more varied and rambling song.

Similar Species
Breeding Bay-breasted Warbler larger, with dark face and throat and buff neck patch. Ruby-crowned Kinglet resembles immature Chestnut-sided, but much smaller; flicks wings nervously; has white wing bars.

Range
Breeds across s. Canada to Nova Scotia, south through Great Lakes region to n. New Jersey, and in Appalachians to n. Georgia. Winters in Central America.

Bay-breasted Warbler
Dendroica castanea

Sparrow-sized
Boreal Forest; Other forest
habitats

Field Marks
5½". A fairly common, dark, chunky warbler occurring in open boreal
forests of spruce and balsam fir. Forages sluggishly in the centers and
tops of dense conifers; its high-pitched song is the best indication of
its presence. Bay-breasteds migrate early in fall, often moving by
mid-August.
Breeding males (1, 2) have black face and chestnut crown, throat, and
sides; underparts and neck patch buff-white; upperparts greenish-
gray, streaked with black; two white wing bars. Breeding females
paler, less chestnut, with a whitish throat and reduced buff on neck.
Winter adults and immatures (3) greener above than in summer;
unstreaked buff-white below with buff undertail coverts and often a
trace of chestnut on flanks. Legs and feet dark in all plumages.
Song a short, high-pitched *seetzy seetzy seetzy*.

Similar Species
Breeding Chestnut-sided Warbler has yellow crown and white throat.
Immature Blackpoll Warbler yellow-green below with faint side
streaks, white undertail coverts, and straw-colored legs. Immature
Pine Warbler grayer or browner with unstreaked back.

Range
Breeds from British Columbia east to central Manitoba and south to
ne. Minnesota, n. New York, and n. New England. Winters in
Central and South America.

Blackpoll Warbler
Dendroica striata

Sparrow-sized
Boreal Forest; Other forest
habitats

Field Marks
5½". A thin-billed, sparrow-sized bird that nests in northern boreal
forests. Most familiar as a very abundant late spring migrant, when it
forages high in tall trees, especially birches and oaks.
Breeding males (1, 2) have distinctive black cap with contrasting
white cheeks bordered by black below; bold black streaks on white
breast; gray back streaked with black; wing bars white; legs
yellowish-pink. Breeding females lack head pattern of male; dark-
streaked gray above (including crown), gray-streaked white below,
with white wing bars. Fall birds (3) resemble breeding female with
strong yellow tinge on head, back, and breast; fall adult males more
darkly streaked than females and immatures. All retain white wing
bars but have darker legs than spring males.
Song highest-pitched of all warbler songs, a series of very thin notes:
seet-seet-seet-seet-seet.

Similar Species
See fall Bay-breasted and immature Pine warblers. Chickadees also
have black-and-white head pattern, but lack streaks and have
different shape. Black-and-white Warbler has streaked crown, bark-
creeping behavior.

Range
Breeds from n. Alaska east to Labrador and Newfoundland, south to
s. Alaska and ne. United States. Winters in South America.

Cerulean Warbler
Dendroica cerulea

Sparrow-sized
Eastern Deciduous Forest

Field Marks

4¾". A small, vocal, short-tailed warbler of mature forests in the East; especially partial to bottomland, swampy areas, and streamside forest. Spends much time high in the canopy, and can be difficult to see; its song is a good clue to its presence. Fairly uncommon, but can be locally abundant.

Adult males (2) light blue to blue-gray above, white below, with dark band across breast and some dark streaks along sides; wings have two white wing bars; white on throat extends up behind ear. Adult females (3) have more gray-blue head and upper back, merging into blue-green to olive on the lower back; duller than male below; two white wing bars. Immatures (1) resemble females, but may be more greenish.

Song is a speedy series of three to five buzzy notes: *zray, zray, zray, zreee.*

Similar Species

Adult male Black-throated Blue Warbler darker blue above, with black face, throat, breast, and flanks. Female Black-throated Blue lacks wing bars; deeper buff below, with no streaks.

Range

Breeds from se. Minnesota to s. New York and s. New England, south to ne. Texas, n. Georgia, and North Carolina. Winters in South America.

Black-throated Blue Warbler
Dendroica caerulescens

Sparrow-sized
Eastern Deciduous Forest

Field Marks
5¼". A small, dark, common warbler of beech- and maple-covered
slopes; usually forages in the middle and lower story of hardwood
thickets, especially where mountain laurel or American yew is
common.

Adult males (1, 2) grayish-blue above with black face, throat, breast,
and sides; belly white; white spot on wing. Females (3) are uniformly
olive-brown above with brownish cheeks and olive-yellow underparts;
the wing patch in females may sometimes be obscured. Immatures
similar to adults.

Song a husky, drawled *zray zray zray zreee*, with slight rising
inflection at the end. Note a dull *smack*.

Similar Species
Dark cheek, white wing spot, and thin bill distinguish female from
stouter-billed and yellower Philadelphia Vireo, and from greener
Tennessee Warbler. See also Cerulean Warbler.

Range
Breeds from south-central Canada east to Nova Scotia, south to Great
Lakes region, w. Pennsylvania, and s. New England; in Appalachians
to n. Georgia. Winters mainly in the West Indies.

Blue-gray Gnatcatcher
Polioptila caerulea

Very Small
Forest habitats; Residential Areas
and Parks

Field Marks
4½". A slender, very active bird with a long tail held straight down
or cocked in a variety of angles. Inhabits open woodland and dense
shrubbery over much of the southern United States. Often utters
buzzy *zhee* notes when foraging.
Adults (1, 2) distinctively blue-gray above, grayish-white below; tail
black above with white edges, mostly white below; bill long and
slender; face shows thin white eye-ring. Breeding males have narrow
black forehead.
Song a mixture of thin notes followed by richer warbles. Call a
nasal *zhee*.

Similar Species
Blue-gray color sets it off from most other species; similarly colored
warblers lack long, thin bill and long, expressively held tail.

Range
Breeds from n. California east across central Great Plains to Great
Lakes region and s. Maine, south through most of United States.
Winters from southernmost parts of range to Mexico.

456

Connecticut Warbler
Oporornis agilis

Sparrow-sized
Northern Wooded Swamps; Boreal
Forest; Thickets

Field Marks
5¾″. An inconspicuous inhabitant of northern wooded swamps; in migration, occurs in damp thickets and wet forested areas. Walks along low branches and ground to feed. Shy; but male's frequent singing calls attention to it.

Adult males (1) have gray hood with conspicuous, complete white eye-ring; upperparts olive-green; dull yellow below. Females (2) similar but duller, with brownish hood. Immatures resemble females; eye-ring may be buff-colored in females and young birds.

Song a loud, clear *beecher, beecher, beecher*. Call a loud, sharp *cheep.*

Similar Species
Mourning Warbler very similar, but lacks eye-ring, or has very faint, incomplete marking about eye; brighter yellow plumage above and below. See also Nashville Warbler.

Range
Breeds from nw. British Columbia east to Quebec and south to northern Great Lakes region. Winters in South America. In spring, migrates through Mississippi Valley; in fall, along Atlantic Coast.

Mourning Warbler
Oporornis philadelphia

Sparrow-sized
Boreal Forest; Northern Wooded
Swamps; Thickets

Field Marks
5¾". A somewhat elusive inhabitant of bogs and wet woods, damp
thickets, and wooded swamps; presence indicated most often by song.
Unlike closely related Connecticut Warbler, this species hops through
undergrowth to forage.

Adult males (1) have dark gray hood with black patch on upper breast;
upperparts olive-green, underparts yellow. Females (2) similar with
reduced and paler hood; lack black patch on throat. Immatures have
gray hood reduced further; some young males may show traces of
black throat patch. Legs and feet pinkish in all plumages. Eye-ring
absent or very faint and incomplete.

Song a loud, liquid *chirry chirry chirry chorry chorry*; call note a loud
chek.

Similar Species
See Connecticut and Nashville warblers. Female and immature
Common Yellowthroats have warm patch of yellow on upper breast,
but remainder of underparts whitish or very pale gray.

Range
Breeds from Rocky Mountains east across Canada to Newfoundland,
south to North Dakota, Great Lakes region, and w. New England; to
Virginia in Appalachians. Winters in Central America.

Nashville Warbler
Vermivora ruficapilla

Sparrow-sized
Boreal Forest; Other forest
habitats

Field Marks
4¾″. A small, ground-nesting summer resident of burnt-over areas, second-growth woods, and thickets in boreal forests; often in groves of aspen or birch. Forages high in trees and sings frequently.
Adults (1, 2) of both sexes have prominent white eye-ring, gray crown and sides of head, olive upperparts, and yellow throat and underparts. Breeding males brighter than females. First-fall birds (3) less clean-cut, with brownish-gray head. Eastern birds tend to be more yellow below; western birds less so, with white belly.
Song a high-pitched, trilled pattern of single or double notes, sometimes ascending the scale.

Similar Species
Breeding male Mourning and Connnecticut warblers have entire head gray; other plumages have traces of the gray hood; all have pink legs. Immature male Common Yellowthroat may have white eye-ring, but lacks contrast of back and crown.

Range
Breeds from s. British Columbia across s. Canada to Nova Scotia, south to central California, n. Utah, Great Lakes region, and n. West Virginia. Winters from California, s. Texas, and s. Florida south to Central America.

Northern Parula
Parula americana

Very Small
Forest habitats

Field Marks

4½". A tiny warbler of the canopy in coniferous forests and mixed
woods; usually found near water. Breeds in a wide range of habitats.
Particularly characteristic of southeastern cypress swamps and boreal
spruce forests, both of which provide mossy nesting materials (lichen
in the North, Spanish moss in the South).

Males (2, 3) blue-gray above with white wing bars and broken eye-
ring; green patch in middle of back; yellow on throat and breast, with
double breast band of black and rusty orange; rest of underparts
white. Females (1), immatures similar but somewhat duller, without
breast bands.

Song a distinctive, buzzy trill, rising in pitch and ending in a sharp
note: *zeee-e-e-e-e-e-ip.*

Similar Species

Tropical Parula, *Parula pitiayumi*, summer visitor to streamside
forest in southern Texas, has dark mask and solid yellow underparts
without dark breast band.

Range

Breeds from se. Manitoba to Nova Scotia, south locally to central
Texas, Gulf Coast, and central Florida; occasionally farther west.
Winters in s. Florida and rarely along Gulf Coast; also in Mexico and
Central America.

460

Golden-winged Warbler
Vermivora chrysoptera

Sparrow-sized
Brushy Open Country; Eastern
Deciduous Forest

Field Marks
4¾". A locally uncommon warbler of open second growth, woodland edges, and brushy pastures; declining in some areas, apparently due to competition with Blue-winged Warbler. Actively forages high and low in trees; sings persistently at height of breeding season.
Adult males (1) have yellow crown, gray upperparts, and yellow wing patches; underparts white, with black throat and ear patch; white spots in tail. Females (2, 3) and immatures have gray throat and ear patch.
Song typically a buzzy, insectlike *beee bz bz bz*; sometimes sings song of Blue-winged Warbler.

Similar Species
"Lawrence's" Warbler, rare hybrid between Golden-winged and Blue-winged warblers, has black mask and throat patch, but underparts bright yellow, wing bars usually white.

Range
Breeds from central Minnesota through Great Lakes region to s. New England, south in Appalachians to n. Georgia. Winters in tropics.

Common Yellowthroat
Geothlypis trichas

Sparrow-sized
Freshwater Marshes; Brushy
Open Country; Thickets

Field Marks
5″. A small, common skulker of thickets, low bushes, tall grass, and
marshes. More often heard than seen, but easily lured into view by
"spishing" noises.
Adult males (1, bottom) have yellow throat and breast, olive-green
upperparts, and bold black mask bordered above with white; belly
white. Adult females (1, top) drab and nondescript but with yellow
throat and undertail, very narrow, pale eye-ring, and diffuse pale
eyebrow. Immature males have much reduced and less well-defined
dark mask and fairly prominent white eye-ring. Immature females
very drab, with buff underparts only tinged yellow on throat and
undertail.
Song a variable, lively warble made up of repeated phrases: *chi-chi-*
wheerchi-chi-wheer chi-chi-wheer or *ter-weechy ter-weechy ter-weechy.*

Similar Species
Kentucky Warbler has differently shaped mask with no white border,
bold yellow spectacles, and entirely yellow underparts. Female and
immature Nashville warblers have more prominent eye-ring, gray
crown. Yellow-breasted Chat much larger with white spectacles.

Range
Breeds from se. Alaska across Canada, south throughout United
States to Mexico. Winters from n. California and New Jersey south
along coasts and from s. United States to Mexico.

Kentucky Warbler
Oporornis formosus

Sparrow-sized
Eastern Deciduous Forest

Field Marks
5¼". A rather secretive inhabitant of wooded ravines and forests,
especially in areas near watercourses. Usually heard before it is seen;
stays on or near the ground to forage for insects, but may move up in
vegetation to sing.
Adult males (1, top) have bold face pattern with black crown, bright
yellow eyebrow, and black patch over ears extending to base of bill;
back and wings uniform olive-green; underparts bright yellow.
Females (1, bottom) similar but with less black on face and head.
Immatures resemble females but may be more brownish on back
and wings.
Song a repeated series of two-syllabled *tur-dle* notes. Call note a
loud *chuck*.

Similar Species
Common Yellowthroat has white on belly; male has black mask.
Female Mourning and Connecticut warblers lack bold black and
yellow pattern on face.

Range
Breeds from se. Nebraska east to se. New York, south to e. Texas,
central Georgia, and North Carolina. Occasionally farther west.
Winters from Mexico to n. South America.

Yellow-breasted Chat
Icteria virens

Sparrow-sized
Brushy Open Country; Thickets

Field Marks
7½". A medium-sized songbird, related to the warblers. Highly vocal, but often difficult to see as it skulks in dense, brushy forest edges and thickets in open country. Usually solitary. Frequently performs a fluttering, leg-dangling display as it sings.

Adults (1–3) plain olive-green above, with blackish lores and dark cheeks set off by white spectacles and mustache streak; throat and breast bright yellow; belly white; flanks tinged with rust; bill black, stouter than in other warblers.

Song a varied medley of clucks, whistles, squawks, and longer phrases; different parts of song so dissimilar that it may be hard to tell which ones are coming from chat, which from other species.

Similar Species
Male Common Yellowthroat smaller, with bold black mask and slender bill.

Range
Breeds from s. British Columbia east across n. United States to Massachusetts, south to s. California, the Gulf Coast, and n. Florida. Winters mainly in tropics, in small numbers along southern Atlantic and Gulf coasts.

Solitary Vireo
Vireo solitarius

Sparrow-sized
Forest habitats

Field Marks
5½". A sparrow-sized songbird with moderately thick bill. Dwells high in canopy of cool hardwood and mixed hardwood-conifer forests. A common spring and fall migrant in most of East.

Adults (1–3) gray on top and sides of head, with bold white spectacles. Back olive-green; wings and tail dusky with pale yellow wing bars and feather edgings; underparts basically white, with greenish-yellow sides and flanks.

Loud song of short, varied phrases: *see me, hear me, here I am*; sweeter, higher-pitched, and with longer pauses than similar song of Red-eyed Vireo.

Similar Species
White-eyed Vireo smaller, with yellow spectacles. Yellow-throated Vireo has yellow throat and spectacles. Female Black-capped Vireo smaller, with darker slate-gray head.

Range
Breeds from British Columbia east across central Canada to Newfoundland, south to North Dakota, Great Lakes region, New England, and in the Appalachians to North Carolina. Winters from s. California and Arizona to Central America, in the East from South Carolina southward.

Black-capped Vireo
Vireo atricapillus

Very Small
Texas: Dry Western Woodlands

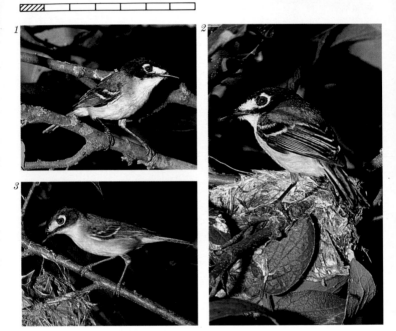

Field Marks
4½". A small, distinctive vireo of brushy oak juniper woodlands.
Forages slowly and deliberately for insects; seems to prefer dry
ravines and hillsides. Sometimes in small flocks at other times than
breeding season.
Adult males (1, 2) have black top and sides of head sharply outlining
broad white spectacles; upperparts bright olive-green with two
yellowish wing bars; underparts white, flanks tinged with yellow; eyes
red. Females (3) similar but with black of head replaced by dark slate-
gray. Immatures like female but browner, with buff underparts.
Song a quick jumble of short, lively chirps, whistles, and twitters.

Similar Species
Solitary Vireo larger, with gray or blue-gray head. White-eyed Vireo
paler with yellow spectacles, white eyes.

Range
Breeds locally in central Oklahoma and w. and central Texas. Winters
in Mexico.

466

White-eyed Vireo
Vireo griseus

Sparrow-sized
Eastern Deciduous Forest;
Thickets

Field Marks
5″. The common small vireo of southeastern thickets, brambles, and viny tangles in the edges of woods and fields. Usually found near the ground. Sings repeatedly and at great length.

Adults (1, 2) mostly pale olive above and white below, tinged with yellow on back, sides, and flanks. Yellow spectacles and two pale yellow or white wing bars are most prominent markings. Eyes white in adults, darker in juveniles.

Lively, loud song begins and ends with sharp, punctuating notes. Variable, but typically: *pick up the beer check!*

Similar Species
Yellow-throated Vireo larger, yellower above and on throat, with dark eyes. Solitary Vireo has white spectacles. Bell's Vireo less boldly marked with narrow spectacles and (usually) a single wing bar.

Range
Breeds from se. Nebraska east to e. Massachusetts, south to central Texas and Florida. Winters from s. Texas along Gulf Coast to Florida and north to coastal North Carolina; also Central America.

Tennessee Warbler
Vermivora peregrina

Sparrow-sized
Boreal Forest; Other forest
habitats

Field Marks
4¾". A small, rather plain warbler that nests in boreal and northern
hardwood forests, and inhabits a variety of wooded habitats in
migration. Forages for insects high in the canopy where it may be
difficult to see; its presence is often signalled by its song.
Breeding males (2) green above, white below, with gray crown, white
eyebrow, and dark gray streak through eye. Females (3) similar but
with yellowish-tinged breast, greenish markings on head. Undertail
coverts usually whitish in all plumages. Fall birds (1) suffused with
greenish-yellow throughout, lack contrasting gray crown.
Song a string of high, thin notes descending the scale and speeding up
into a trill at the end.

Similar Species
Orange-crowned Warbler dingier green overall, with less prominent
eyebrow and yellow undertail coverts. Warbling and Philadelphia
vireos have similar color pattern, but are heavier-billed and slower-
moving. Warbling Vireo less green above; Philadelphia yellower
below.

Range
Breeds from se. Alaska across Canada to Newfoundland, south to
British Columbia, Great Lakes region, and s. Maine. Winters mainly
in Mexico, occasionally s. California and s. Texas.

Orange-crowned Warbler
Vermivora celata

Sparrow-sized
Boreal Forest; Other forest
habitats

1

Field Marks
5″. A very drab, olive-gray warbler found in open woods, edges, and thickets. Nests in the Far North, and more commonly in the West than in the East; but a familiar winter resident in the Southeast. Shows considerable geographic variation in color.
Adults (1) dull olive-green (yellower in the West, grayer in the East), with a vague, pale eyebrow and faint diffuse streaks below. Females similar but grayer. Both sexes slightly paler below than above, with pale yellow undertail coverts. Orange crown patch of male rarely visible in the field.
Song a trill of short single notes that may change speed and pitch.

Similar Species
See Tennessee and Wilson's warblers.

Range
Breeds from Alaska across Canada to Labrador, south to Baja California and w. Texas. Winters from n. California, central Arizona, Texas, and se. Virginia south to Central America.

Yellow-throated Vireo
Vireo flavifrons

Sparrow-sized
Eastern Deciduous Forest

1

Field Marks
5½". A sparrow-sized bird of the high canopy of mature broad-leaved deciduous forests. Less often encountered than most other eastern vireos. Found mainly along forest edges, near streams, and in parks where tall hardwoods occur near open areas. Usually sings from high in trees.

Adults (1) moderately thick-billed, with yellow spectacles, throat, and breast; crown and back olive; rump gray; belly and wing bars white. Immatures resemble adults.

Song a series of short, widely spaced two- and three-syllabled whistled phrases, similar in pattern to that of Red-eyed Vireo but louder, often with a burry quality.

Similar Species
Pine Warbler has narrower yellow spectacles, thinner bill, yellowish rump, and faint streaks on breast and flanks. Yellow-breasted Chat lacks wing bars, has white spectacles; rarely seen in forest canopy.

Range
Breeds from s. Manitoba and Minnesota east to Maine, south to e. Texas, Gulf Coast, and n. Florida. Winters in Central and South America, in small numbers in Florida.

Philadelphia Vireo
Vireo philadelphicus

Sparrow-sized
Forest habitats

Field Marks
5¼". A rather uncommon small vireo found nesting in second growth of clearings, burns, and streamsides, and in a variety of forest habitats as a migrant. The least common vireo in the East, less vocal than most other vireos in migration.

Breeding adults (1) have gray crown, white eyebrow, dark stripe through eye, olive upperparts, and dingy white underparts tinged with yellow on breast and sides. May show single buff wing bar. Fall adults mostly greenish-yellow below. Immatures resemble adults. Song closely resembles that of Red-eyed Vireo but is somewhat higher, thinner, and slower.

Similar Species
Warbling Vireo paler overall, with fainter eyebrow and eyestripe, whiter underparts. Tennessee Warbler more crisply marked, with greener back, whiter underparts, more contrasting gray crown, and thinner bill.

Range
Breeds from s. British Columbia across Canada to Newfoundland, south to Great Lakes region and n. New England. Winters from Mexico to South America.

Swainson's Warbler
Limnothlypis swainsonii

Sparrow-sized
Southern Wooded Swamps; Moist
Thickets

Field Marks
5½″. A shy, rather nondescript warbler that lives in dense thickets, canebrakes, and wooded swamps in the Southeast. Spends much time close to the ground, but gives away its presence in the vegetation with its loud, ringing song.

Adults (1) drab olive-brown above with reddish-brown cap and pale eyebrow; dark line runs through eye to nape; underparts pale yellowish- or greenish-white; bill long and tapering. Immatures resemble adults.

Song consists of three or four loud notes, followed by a jumble of falling notes: *teeu-teeu-teeu, witch-er-you-teeu*; reminiscent of song of more common Louisiana Waterthrush.

Similar Species
Philadelphia Vireo paler, with yellow tinge on throat and breast, grayish-olive upperparts and crown. Red-eyed Vireo greener, with bold eyebrow, gray crown; whitish underparts. Worm-eating Warbler has bold head stripes, lacks rufous cap.

Range
Breeds from s. Arkansas, s. Indiana, West Virginia, and Delmarva Peninsula south to e. Texas, Gulf Coast, and n. Florida. Winters in tropics.

Red-eyed Vireo
Vireo olivaceus

Sparrow-sized
Forest habitats; Groves

Field Marks
6″. One of North America's most abundant songbirds, found in a variety of mixed broad-leaved forests and in groves, parks, and residential areas. Rather sluggish in movements, and often difficult to see among foliage, but its cheerful and persistent song is a good clue to its presence.
Adults (1) olive-green above, dull white below, with gray crown; white eyebrow bordered above and below with dark stripes; eyes red. First-fall birds have brownish eyes and are tinged yellow below.
Song a series of short, widely spaced, varied phrases: *see me, hear me, see you, vireo,* sung continuously throughout the day.

Similar Species
Warbling Vireo lacks gray crown; has grayer back, duller eyebrow; song different.

Range
Breeds from British Columbia and s. Mackenzie east to Maritime Provinces, south to n. Oregon, the Great Plains, the Gulf Coast, and central Florida. Winters in South America.

Black-whiskered Vireo
Vireo altiloquus

Sparrow-sized
Southern Florida: Mangroves

Field Marks
6¼″. A summer visitor to coastal mangroves and hammocks in
Florida, sometimes found in willows along the northern Gulf Coast.
May sing all day long, like the closely-related Red-eyed Vireo.
Adults (1, 2) olive-green above, dull white below, with gray crown,
buff-white eyebrow, dark gray eyeline, and thin dark-gray whisker
mark; bill fairly large; eyes red. Immatures more brownish above,
buffier below, with pale yellow wing bar and brown eyes.
Song a series of short phrases; sometimes indistinguishable from that
of the similar Red-eyed Vireo, but may be more forceful and have
doubled phrases.

Similar Species
Red-eyed Vireo has smaller bill, lacks dark whisker mark.

Range
Breeds along Gulf Coast of Florida to the Keys. Winters in South
America.

Warbling Vireo
Vireo gilvus

Sparrow-sized
Forest habitats; Groves; Open
Country; Residential Areas and
Parks

Field Marks
5½″. A widespread and abundant inhabitant of open forested areas, woodland groves, parks, and suburbs; partial to tall streamside trees. Rather nondescript, but, like most vireos, sings persistently. Forages sluggishly high in trees.
Adults (1, 2) olive-gray above, white below, with a diffuse pale eyebrow; crown gray, not strongly contrasting with back.
Song a long, continuous warble, like Red-eyed Vireo's song with the pauses removed.

Similar Species
Philadelphia Vireo has yellowish underparts, dusky lores.

Range
Breeds from n. British Columbia across s. Canada, south to Mexico, n. Alabama, and Virginia. Winters in Mexico and Central America.

Bell's Vireo
Vireo bellii

Sparrow-sized
Streamside Forest; Thickets

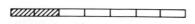

1

Field Marks
4¾". A small, chunky, drab vireo of streamside willows, thickets, and groves. More energetic than other vireos, actively flitting about and singing all day. Great geographical variation in color, face pattern, and number of wing bars.
Adults (1) usually olive-green above, yellowish below, with pale spectacles, eyebrow, and one or two pale wing bars. Immatures similar to adults.
Song a distinctive jumble of chattery, harsh notes, becoming louder toward the end.

Similar Species
Warbling Vireo has pale eyebrow rather than spectacles. White-eyed Vireo has two wing bars, yellow spectacles. All three have very different songs. The Northern Beardless-Tyrannulet is smaller, with gray wash on breast and buff wing bars; usually raises feathers of crown to create crested look; call a whining *pee yee*. See Ruby-crowned Kinglet.

Range
Breeds from s. California (rare), s. Nevada, New Mexico, w. Texas, North Dakota, Illinois, and Louisiana south into Mexico. Winters in Mexico and Central America.

Ruby-crowned Kinglet
Regulus calendula

Very Small
Boreal Forest; Eastern Deciduous
Forest; Residential Areas and
Parks

Field Marks
4¼″. A tiny, short-tailed bird, often seen in loose flocks with
chickadees, nuthatches, or other small birds. Breeds in coniferous
forest of the Far North and western mountains. Flight is typically
weak and usually short. Kinglets nervously flick their wings almost
continuously.
Adults (1) olive-gray above, whitish-buff below, with white eye-ring;
two white wing bars, leading one frequently concealed, black band
below second; bill fine, pointed, and black. Adult male's bright red
crown patch often concealed.
Call a hard, quick *ji-dit*. Song a few whistles, giving way to a warble,
tee tee tew tew tideedadee.

Similar Species
Golden-crowned Kinglet has white eyebrow and line below eye, black
crown with yellow to yellow-orange center, bolder white wing bar,
grayer upperparts; more closely associated with conifers; call a high,
thin *see see see see*; song has same notes but rises in pitch and ends in
chatter. See Bell's Vireo.

Range
Breeds from nw. Alaska across Canada to Newfoundland, south to
Baja California and New Mexico, Great Lakes region, and n. New
England. Winters from British Columbia, n. Texas, and Maryland
south.

Appendices

Credits

The first number indicates the page on which photographs appear. The position of the photograph on the page is indicated in parentheses; c indicates the circle inset, and if a number in parentheses is followed by the letter L or R, this number-and-letter combination indicates the row and position within the row, left (L) or right (R). Some photographers have photographs under agency names as well as their own.

Photographers

David G. Ainley: 37(1), 48(2), 55(1), 56(2).

B. T. Aniskowicz: 138(1).

Robert H. Armstrong: 2(4L), 5(3L,4L), 79(3), 97(2,3), 113(C), 141(C), 321(C), 379(1), 477(C).

Ron Austing: 155(1), 187(2), 223(1), 224(C), 231(4), 235(2), 236(1), 237(1), 240(1), 241(1,2), 242(C), 243(C), 246(1), 261(1), 270(1), 279(C), 284(C,1,2), 298(2), 311(C), 319(1), 331(1), 352(C,3), 378(C,2), 383(1,2), 395(C,1), 396(1), 400(1,2), 414(1), 416(1), 428(1), 431(1), 447(C,2), 456(C,1,2).

Stephen F. Bailey: 3(4L), 6(2R), 11(2L), 59(6), 150(C), 151(C), 197(1), 217(3), 230(1), 314(2), 442(2), 474(1).

Kathleen Blanchard: 65(C,1).

William J. Bolte: 21(1,3), 32(2), 38(C,4), 74(1), 129(1), 172(1), 296(2), 317(C), 375(1).

L. Page Brown: 110(1), 173(1), 228(C,2), 268(2), 474(C,2).

Fred Bruemmer: 5(2R,4R), 24(1), 50(C,1,2), 80(1), 81(C), 169(2), 411(1).

N. R. Christensen: 164(1), 193(1).

Roger B. Clapp: 6(1R), 12(2L), 21(2), 53(1), 58(1,3), 81(2), 214(1).

Herbert Clarke: 23(2), 42(3), 61(1), 74(3), 78(1,2), 79(2), 88(2), 100(1), 113(3), 121(3), 185(2), 224(3), 241(C), 272(1), 287(C,1), 302(2), 309(1), 351(3), 357(C), 398(C,2,3), 399(C,1), 412(C,1), 469(C,1).

Bruce Coleman, Inc.: E. Duscher, 424(C,1); Edgar T. Jones, 408(C,1).

Cornell Laboratory of Ornithology: L. Page Brown, 53(C), 239(1), 296(1); Betty Darling Cottrille, 414(2), 457(2), 461(3); John Henry Dick, 303(1); John S. Dunning, 390(3), 433(C); Bill Dyer, 333(1), 455(C,2), 460(2), 468(1,3); Lang Elliott, 267(1,2); Grimes, 464(1);

Index

Companion Volumes
Audubon Handbooks

How to Identify Birds
Truly a breakthrough in field
identification, this handbook uses
clues a person can easily see and
remember to solve the mysteries
of identification. It shows step-by-
step how to identify birds in the
field. Specially keyed to *Western
Birds* and *Eastern Birds*, the
volume contains more than 700
color photographs, 33 unique
charts of field marks, maps, and
a clear, practical text.

Western Birds
This practical guide covers every
breeding species west of the
Rockies, with 1,314 color
photographs and 168 drawings.
Picture-and-text accounts use the
easy-to-remember field marks
explained in *How to Identify
Birds*. Twelve special pages of
flight comparison photographs for
swans, geese, ducks, raptors, and
gulls.